Intervention Workshop
Progress in
Mathematics

SADLIER-OXFORD

Rose Anita McDonnell

Catherine D. LeTourneau

Anne Veronica Burrows

Anne Brigid Gallagher

Francis H. Murphy

M. Winifred Kelly

with
Dr. Elinor R. Ford

Sadlier-Oxford
A Division of William H. Sadlier, Inc.

Table of Contents

Introduction

Section I – *Intervention Workshop* Lessons

Section II – Student Pages

Titles of Student Lessons

Section III – Blackline Masters

Titles of Blackline Masters

Blackline Master Page

Foreword

Throughout the United States rigorous standards in mathematics are being established. In order to achieve these standards, students must have a solid understanding of basic mathematical concepts and skills. Many students, however, do not have such an understanding. The goal of *Intervention Workshop* is to provide a fully integrated system of remediation that includes materials for reteaching at-risk students. Each lesson presents essential concepts or fundamental skills that students may not have previously mastered. When these concepts and skills have been mastered, students will have the understanding and background necessary to succeed at current and future grade-level standards.

The following are the six steps of *Intervention Workshop*.

1. Identify textbook lessons with precursor skills. Throughout Grade 6, certain lessons and specific exercise sets have been identified by the authors as representing a *precursor skill*, mastery of which is necessary if students are to meet rigorous grade-level expectations. (In *Progress in Mathematics* Teacher's Edition these exercises are highlighted in red and are identified by the *Intervention Workshop* logo next to them.) A preliminary look at the *Key to Using Progress in Mathematics and Intervention Workshop* on pages 2–5 of this book will provide an overview of the lessons identified as essential, if growth in mathematics is to occur.

2. Use observational assessment to determine competency. When you begin a lesson in the Teacher's Edition of *Progress in Mathematics* that has an exercise set in red with the *Intervention* logo, note whether any student is experiencing difficulty with the indicated exercises. If a student is having problems with the material, *Intervention Workshop* is recommended.

3. Use *Keys* to identify the appropriate *Intervention* lesson. The *Key to Using Progress in Mathematics and Intervention Workshop* (see pages 2–5) or the *Key to Intervention Workshop* on the H page in each *Chapter Overview* in the Teacher's Edition of *Progress in Mathematics* explains how *Intervention Workshop* lessons correlate to the student text. Both *Keys* also explain how the *Skills Update* lessons in *Progress* can be used in conjunction with *Intervention Workshop*.

4. Determine the student's needs. After identifying the appropriate intervention, you or a tutor should use the *Diagnostic Interview* questions at the beginning of each lesson to verify the problem a student is having. The interview should also help you focus on the precursor skills and concepts that are to be remediated.

5. Present the lesson. Read through the stepped-out, scripted presentation of either the prescribed interactive student lesson (see pages IW1–IW32 in Section II of this book) or the Teacher's Edition-only lesson with the students. (For each Teacher Edition-only section, there are three or four activities available. The *Background* section and the *Diagnostic Interview* are designed to help you choose which activities are the most appropriate.)

6. Check learning. After working through the presentation with students, you or a tutor can check students' understanding by administering the *Summing-It-Up Quiz*. This mini-assessment helps you determine if students are ready to return to the Grade 6 lesson in the textbook and to continue working toward the grade-level standards.

Intervention Workshop is not meant to replace or become a parallel program to Grade 6 *Progress in Mathematics*. A speedy return of the student to the classroom and the curriculum is expected.

Key to Using *Progress in Mathematics*

Lesson Number	*Progress in Mathematics* Lesson Title	Textbook Pages	Key Exercises	*Skills Update* Reference
	Progress in Mathematics			
1-1	How Much is a Trillion?	38–39	6–7	Use Numeration I before *IW.*
2-1	Properties of Multiplication	64–65	9–22	
2-3	Estimating and Finding Products	68–69	1–5	Use Operations V then VI after *IW.*
2-6	Short Division and Divisibility	74–75	1–12	
2-7	Estimating Quotients	76–77	1–12	Use Operations VII then VIII after *IW.*
3-1	Decimals	94–95	1–11	Use Numeration VI after *IW.*
3-4	Compare and Order Decimals	100–101	1–19	
4-3	Multiplying Decimals by Whole Numbers	126–127	6–23	
4-5	More Multiplying Decimals	130–131	1–15	
4-9	Dividing Decimals by Whole Numbers	138–139	3–18	
4-10	Dividing by a Decimal	140–141	5–32	
5-4	Prime Factorization	168–169	12–21	Use Numeration IV "Factors" after *IW.*
5-6	Fractions in Simplest Form	172–173	25–48	
6-3	Adding Fractions	207–208	1–17	Use Fractions I then II after *IW.*
6-4	Adding Mixed Numbers	208–209	1–9	
6-5	Subtracting Fractions	210–211	13–16	Use Fractions I then II after *IW.*
6-6	Subtracting Mixed Numbers	212–213	17–36	
7-1	Multiplying Fractions by Fractions	226–227	3–17	
7-2	Multiplying Fractions and Whole Numbers	228–229	1–24	
7-4	Multiplying Mixed Numbers	232–233	9–31	
7-9	Dividing a Mixed Number	242–243	3–34	
8-2	Surveys	260–261	8–12	
8-4	Range, Mean, Median, and Mode	264–265	1–8	

and *Intervention Workshop*

	Progress in Mathematics			
Lesson Number	*Progress in Mathematics* Lesson Title	Textbook Pages	Key Exercises	*Skills Update* Reference
8-7	Making Line Graphs	270–271	10–13	Use Statistics and Graphs before *IW*.
8-11	Probability	278–279	23–26	
8-12	Compound Events	280–281	15–18	
9-3	Measuring and Drawing Angles	300–301	7–18	
9-7	Classifying Triangles	308–309	9–12	
9-8	Classifying Quadrilaterals	310–311	1–7	
10-9	Using Perimeter	348–349	1–10	Use Measurement VI after *IW*.
10-12	Area of Triangles and Parallelograms	354–355	1–11	Use Measurement VI after *IW*.
11-2	Equal Ratios	378–379	13–24	
11-4	Proportions	382–383	1–8	
11-5	Solving Proportions	384–385	3–14	
12-2	Percent Sense	414–415	13–24	
12-5	Using Percent to Solve Problems	420–421	1–10	
13-4	Another Addition Model	448–449	3–17	
13-10	Graphing Ordered Pairs of Integers	460–461	1–22	
14-1	Algebraic Expressions	474–475	11–16	
14-4	Solving Equations: Add and Subtract	480–481	1–4	
14-12	Rational Numbers: Number Line	496–497	1–16	

and *Intervention Workshop*

Section I

Intervention Workshop Lessons

Understanding Place Value

Use with Lesson 1-1, exercises 6–7, text pages 38–39.

DIAGNOSTIC INTERVIEW

Before beginning this *Intervention Workshop* lesson, have students complete the following exercises.

Diagnostic Exercises

◆ **Say:** *How are the numbers 610 and 601 the same? How are they different?* (Same: There are 3 digits in each number, or the same number of digits; 6 is in the hundreds place, or starts both numbers. Different: In 610, 1 is in the tens place, or in the middle; 0 is in the ones place, or is the last number. In 601, 1 is in the ones place, or is the last number; 0 is in the tens place, or in the middle.)

◆ **Say:** *Write these numbers. Name the place and the value of the digit 7 in each number.*

    704   172,005   760,005

(From the left: 7 is in the hundreds place, the value is 700; 7 is in the ten thousands place, the value is 70,000; 7 is in the hundred thousands place, the value is 700,000.)

If students answer the Diagnostic Exercises correctly, there is probably no need for them to do this *Intervention Workshop* lesson. Have those students rejoin their class on page 39 of the student textbook. If, however, students have difficulty with the exercises above, you should first assign the review lesson on page 1 of *Skills Update* in the student textbook before proceeding with this *Intervention Workshop* lesson.

BACKGROUND

This *Intervention Workshop* lesson addresses possible difficulties with the relationship between place and value in whole numbers. Students begin by observing digit position and draw upon their intuitive understanding of how sets of two to four small numbers are the same and how they are different. Next students determine the place and value of a given digit by analyzing its position in a number written in a place-value chart. To help students build an understanding of the relationship between place and value, the exercises present progressively larger numbers.

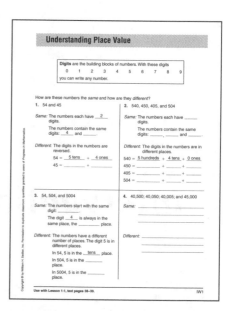

TEACHING THE *INTERVENTION WORKSHOP* LESSON

Getting Started

Materials: centimeter grid paper (BLM 1), place-value chart (BLM 2)

◆ **Say:** *A digit is a number from zero to 9. A number can have just one digit or it can have several digits. You can describe a number by the number of digits that it has: for example, a 1-digit number, a 2-digit number, a 10-digit number.*

◆ **Say:** *Write any 1-digit number.* (sample answer: 5) *Write any 2-digit number.* (sample answer: 23) *Write any 3-digit number.* (sample answer: 416) *Write any 4-digit number.* (sample answer: 3289)

◑ USING PAGES IW1 and IW2

Have a student read aloud the first sentence in the box at the top of page IW1. Relate the given definition to the numbers students wrote earlier in *Getting Started* in the following way.

◆ **Say:** *Name the digits in the numbers you wrote.* (sample answer: 5; 2 and 3; 4, 1, and 6; 3, 2, 8, and 9)

Exercises 1–4 on page IW1

Have a student read aloud the direction line for exercises 1–4. These four exercises lead students to analyze the place and value of the digits 4 and 5 in increasingly greater numbers.

◆ **Say:** *You may want to write the numbers vertically, one number below the other, on grid paper. When you write them this way, you will see how the numbers are the same and how they are different.*

For students who have difficulty with exercise 1 use the following additional example: 68 and 86.

For exercises 2–4, continue to have students work in the same manner and assist them as necessary (additional example for exercise 2: 806, 608, 680, and 860; additional example for exercise 3: 86, 806, 8006).

Before beginning exercise 4 you may want to provide an interim exercise with four numbers in thousands to help students make the transition from exercise 2 to exercise 4: 5004, 4005, 4500, and 5400 (additional example for exercise 4: 86,000; 60,008; 80,600; and 60,080).

Exercises 5–13 on page IW2

Have a student read aloud the material in the box at the top of page IW2.

◆ **Say:** Place *means the position of a digit in a number. When a digit is assigned to a place other than the ones place in a number, it takes on a new value. For example in the number 428, the 4 is in the hundreds place. The value of 4 in the number 428 is 4 times 100, or 400.*

You can explain the place and value of the digits 2 and 8 in the same way.

Assist students as necessary to complete exercises 5–11, in which the numbers are already written in place-value charts. You may need to provide students with additional practice for exercises 12–13 by having them write other numbers in place-value charts and determine the place and value of a given digit.

SUMMING-IT-UP QUIZ

Give this quiz to students.

◆ **Say:** *Write these three numbers. Name the place and value of the digit 8 in each number.*

        780   *18,070*   *870,300*

(From the left: 8 is in the tens place, the value is 80; 8 is in the thousands place, the value is 8000; 8 is in the hundred thousands place, the value is 800,000.)

◆ **Say:** *Look at the three numbers you wrote. Write the number with a digit that has a value of 70,000.* (The number is 870,300 and the digit is 7.)

If students answer the *Summing-It-Up Quiz* correctly, they should return to Lesson 1-1 and resume work on page 39 of their textbook. If students do not successfully complete the *Summing-It-Up Quiz*, further remediation may be necessary.

Answers: Pages IW1 and IW2

1. 5; 4 tens + 5 ones

2. 3; 5, 4, 0;   4 hundreds + 5 tens + 0 ones
                   4 hundreds + 0 tens + 5 ones
                   5 hundreds + 0 tens + 4 ones

3. 5; ones; hundreds; thousands

4. *Same:* The numbers start with the digit 4; the numbers have 5 digits; the digit 4 is always in the ten thousands place.
Different: The digit 5 is in different places; 5 is in the hundreds place of 40,500; 5 is in the tens place of 40,050; 5 is in the ones place of 40,005; 5 is in the thousands place of 45,000.

5. 300

6. ones; 3

7. ones; 1; 3

8. thousands; 3000

9. tens; 30

10. ten thousands; 30,000

11. thousands; 3000

12. 7   3   0   0   0   0
     ten thousands; 30,000

13. 3   0   0   0   7   0
     hundred thousands; 300,000

Meaning of Multiplication

Use with Lesson 2-1, exercises 9–22, text pages 64–65.

DIAGNOSTIC INTERVIEW

Before beginning this *Intervention Workshop* lesson, have students complete the following exercises.

Diagnostic Exercises

Write these two expressions on the board.

$$3 \times 5 \qquad\qquad 5 \times 3$$

◆ **Say:** *How are these expressions alike? How are these expressions different?*
(Possible responses: They are alike in that they involve only multiplication; they involve the same two factors, 3 and 5; and they have a product of 15. They are different in that the factors are in different orders.)

Write these two expressions on the board.

$$(4 \times 2) \times 3 \qquad\qquad 4 \times (2 \times 3)$$

◆ **Say:** *How are these expressions alike? How are these expressions different?*
(Possible responses: They are alike in that they involve only multiplication; they involve the same three factors, 4, 2, and 3; and they have a product of 24. They are different in that they have parentheses around different numbers.)

Write this expression on the board.

$$4 \times a$$

◆ **Say:** *For what value of* a *would the expression 4 times a equal 4? equal zero?* (1; 0)

Write these three expressions on the board.

$$3 \times (6 + 4) \qquad (3 + 6) \times (3 + 4) \qquad (3 \times 6) + (3 \times 4)$$

◆ **Say:** *Which two expressions have the same value? How do you know?*
(The expressions $3 \times (6 + 4)$ and $(3 \times 6) + (3 \times 4)$ have the same value. When a pair of addends in parentheses is multiplied by a number, it is the same as multiplying each addend by the number and adding the two products.)

If students answer the Diagnostic Exercises correctly, there is probably no need for them to do this *Intervention Workshop* lesson. Have those students rejoin their class on page 65 of the student textbook. For students who have had difficulties answering the exercises above, continue the *Intervention Workshop*.

BACKGROUND

Exercises 9–22 on page 65 of the student textbook require students to use the properties of multiplication to complete multiplication exercises. These *Intervention Workshop* activities address the needs of students who still have not assimilated the general meanings and properties of multiplication. In Activity One students use an area model to understand the commutative property of multiplication. In Activity Two students review the identity and zero properties of multiplication. Activity Three provides an opportunity to model the associative property of multiplication. In Activity Four students use a model to understand the distributive property of multiplication over addition. Use one or more of these activities with your students, as needed.

TEACHING THE *INTERVENTION WORKSHOP* LESSON

Getting Started

◆ **Say:** *What does 4 times 3 equal?* (12) *What are some other multiplication sentences that have the same product?* [some possible responses: $1 \times 12 = 12$, $2 \times 6 = 12$, $3 \times 4 = 12$, $6 \times 2 = 12$, $12 \times 1 = 12$; $2 \times (2 \times 3) = 12$, $(2 \times 2) \times 3 = 12$, $3 \times (2 \times 2) = 12$]

ACTIVITY ONE: Use Models to Understand the Commutative Property of Multiplication

Materials: centimeter grid paper (BLM 1)

Have students shade 5 rows of 4 squares each on a piece of grid paper.

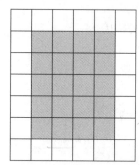

◆ **Say:** *How many rows are there in all?* (5) *How many small squares are shaded in each row?* (4) *How many small squares are shaded in all?* (20) *What multiplication fact does the model show?* ($5 \times 4 = 20$)

Have students record the multiplication fact. Then have students rotate the sheet of grid paper so that the long side of the rectangle is the width and the short side is the height.

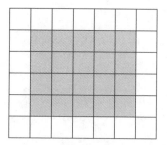

◆ **Say:** *How many rows are there in all?* (4) *How many small squares are shaded in each row?* (5) *How many small squares are shaded in all?* (20) *What multiplication fact does the model show?* ($4 \times 5 = 20$)

Have students compare the two multiplication sentences that they wrote to describe the model.

◆ **Say:** *How are these two multiplication sentences alike? How are they different?* (Possible responses: They are alike in that they involve the same two factors, 4 and 5, and have a product of 20; they are different in that the factors are in different orders.)

Have students repeat the activity independently several times. First students choose the dimensions of a rectangle, draw the rectangle, and write a multiplication fact for the rectangle. They then rotate the rectangle and write a second multiplication fact.

ACTIVITY TWO: Use the Meaning of Multiplication to Understand Multiplying By 0 or 1

Materials: counters

Lead students to derive the identity property of multiplication. Have students show 6 groups with 1 counter in each group.

◆ **Say:** *How many groups of counters are there?* (6) *How many counters are in each group?* (1) *How many counters are there in all?* (6) *What multiplication fact does the model show?* ($6 \times 1 = 6$) *What other multiplication fact can you write that uses the same three numbers?* ($1 \times 6 = 6$)

Have students show 4 groups with 1 counter in each group, 7 groups with 1 counter in each group, and 12 groups with 1 counter in each group. For each model students write two multiplication facts: one for the model and a second fact that uses the same three numbers.

◆ **Say:** *What rule can you use to multiply a number and 1?* (Possible response: When you multiply a number and 1, the product is the factor other than 1.)

Now lead students to derive the zero property of multiplication. Write the following on the board.

$$0 \times 7 = \underline{\ ?\ }$$

◆ **Say:** *How many groups of 7 does the multiplication sentence represent?* (0) *If you have zero groups of 7 counters, how many counters do you have in all?* (0) *What does zero times 7 equal?* (0) *What other multiplication fact can you write that uses the same three numbers?* ($7 \times 0 = 0$)

Write the following on the board.

$$0 \times 8 = \underline{\ ?\ } \quad 0 \times 11 = \underline{\ ?\ } \quad 0 \times 9 = \underline{\ ?\ }$$

Have students complete each example, then write a second multiplication fact using the same three numbers. Let them draw pictures or use counters if they wish.

◆ **Say:** *What rule can you use to multiply a number and zero?* (Possible response: When you multiply a number and zero, the product is zero.)

ACTIVITY THREE: Use Models to Understand the Associative Property of Multiplication

Materials: connecting cubes

Write the following on the board.

$$4 \times (2 \times 5) = \underline{\ ?\ }$$

Remind students that the operation within the parentheses is done first. Have students use connecting cubes to show the operation in the parentheses by making 2 rows of 5 cubes each and connecting them so they form a rectangle.

◆ **Say:** *You have a model that shows 2 times 5. How many of these models do you need to make in all in order to complete the multiplication example?* (4) *How do you know?* (The product of 2 and 5 is being multiplied by 4.)

Have students finish modeling the multiplication by making 3 more 2 × 5 rectangles.

◆ **Say:** *How many cubes did you show in all?* (40)

Write the following on the board.

$$(4 \times 2) \times 5 = \underline{\ ?\ }$$

Have students use connecting cubes to show the operation within the parentheses by making 4 rows of 2 cubes and connecting them so they form a rectangle.

◆ **Say:** *You have a model that shows 4 times 2. How many of these models do you need to make in order to complete the multiplication?* (5) *How do you know?* (The product of 4 and 2 is being multiplied by 5.)

Have students finish modeling the multiplication by making 4 more 4 × 2 rectangles.

◆ **Say:** *How many cubes did you show in all?* (40) *Look at the two multiplication sentences on the board. When we moved the parentheses, we changed the grouping of the numbers we were multiplying. Did changing the grouping of the numbers change the product?* (No)

ACTIVITY FOUR: Use Models to Understand the Distributive Property

Materials: centimeter grid paper (BLM 1), overhead projector, grid transparency

Place a grid transparency on the overhead. Outline a rectangle that has 3 rows of 12 squares each. Have students do the same on their grid paper. Then draw a line down the rectangle dividing it into 2 rectangles: one that has 3 rows of 10 squares each and one that has 3 rows of 2 squares each. Have students do the same. Label the rectangle as shown at the right.

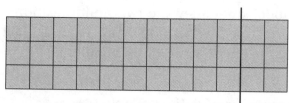

$$3 \times (10 + 2) = \underline{\ ?\ }$$

◆ **Say:** *How does the multiplication sentence represent the picture?* (Possible response: There are 3 rows of squares. Each row has 10 squares + 2 squares, or 12 squares.) *How many squares does the large rectangle have in all?* (36)

Write the product on the overhead. Now point out the two smaller rectangles created by the dividing line.

◆ **Say:** *What multiplication sentence does the rectangle on the left of the line show?* (3×10) *What multiplication sentence does the rectangle on the right of the line show?* (3×2)

Write $(3 \times 10) + (3 \times 2) = \underline{?}$ below the diagram.

◆ **Say:** *How does the multiplication sentence represent the picture?* [Possible response: (3×10) represents the number of squares in the left rectangle. (3×2) represents the number of squares in the right rectangle. To find the total number of squares in the two small rectangles, you add the two amounts: $(3 \times 10) + (3 \times 2).$]

Have students complete the multiplication sentence. Then write the following underneath the sentence $3 \times (10 + 2) = \underline{?}$ below the diagram.

$$3 \times (10 + 2) = (3 \times 10) + (3 \times 2)$$

Write the following exercises on the board. Have students work independently, drawing and labeling rectangles, if helpful, to complete the exercises.

$$5 \times (4 + 3) = (5 \times \underline{?}) + (5 \times 3)$$
$$6 \times (\underline{?} + 4) = (6 \times 5) + (6 \times 4)$$
$$\underline{?} \times (2 + 7) = (3 \times 2) + (3 \times 7)$$

SUMMING-IT-UP QUIZ

Give this quiz to students.

◆ **Say:** *3 times 5 equals 5 times what number?* (3)

Write these two equations on the board or overhead.

$$(3 \times 3) \times 2 = 24 \qquad 3 \times (3 \times 2) = \underline{?}$$

◆ **Say:** *How does knowing the answer to the first problem help you find the answer to the second problem?* (Possible response: Changing the grouping of the factors does not change the product. So the answer to the second problem is the same as the answer to the first problem.)

◆ **Say:** *What happens when you multiply a number and zero?* (The product is 0.) *What happens when you multiply a number by 1?* (The product is the same as the number that is being multiplied by 1.)

Write these four expressions on the board or on the overhead.

a. $4 \times (2 + 3)$ b. $(4 \times 5) + (4 \times 5)$

c. $(4 \times 2) + (4 \times 3)$ d. $(4 \times 2) \times (4 \times 3)$

◆ **Say:** *Which other expression is equal to expression* a? (c)

If students answer the *Summing-It-Up Quiz* correctly, they should then resume work on page 65 of their textbook. If students do not successfully complete the *Summing-It-Up Quiz,* further remediation may be necessary.

Use with Lesson 2-3, exercises 1–5, text pages 68–69.

DIAGNOSTIC INTERVIEW

Before beginning this *Intervention Workshop* lesson, have students complete the following exercises.

Diagnostic Exercises

Write this exercise on the board.

$$
\begin{array}{r}
346 \\
\times\ 49 \\
\hline
3114 \\
13840 \\
\hline
\end{array}
$$

◆ **Say:** *How do you find the partial product 3114?* (Multiply 9 × 346.) *How do you find the partial product 13,840?* (Multiply 40 × 346.) *How do you use the partial products 3114 and 13,840 to find the product for 49 × 346?* (Add the partial products.) *What does 49 times 346 equal?* (16,954)

Write this exercise on the board.

$$
\begin{array}{r}
279 \\
\times\ 527 \\
\end{array}
$$

◆ **Say:** *Find the product for 7 times 279. What is the product?* (1953) *Find the product for 20 times 279. What is the product?* (5580) *Find the product for 500 times 279. What is the product?* (139,500) *What does 527 times 279 equal?* (147,033)

If students answer the Diagnostic Exercises correctly, there is probably no need for them to do this *Intervention Workshop* lesson. But before they resume work on page 69, you may want to assign the review lessons on pages 11 and 12 of *Skills Update* in the student textbook. Then have students rejoin their class on page 69. For students who have had difficulty, continue this *Intervention Workshop*.

BACKGROUND

Exercises 1–5 on page 69 of the student textbook require students to multiply a 3- or 4-digit number by a 3-digit multiplier. These *Intervention Workshop* activities address the needs of students who still have not assimilated the procedure of multiplying multidigit numbers place-by-place or recording each step in the multiplication algorithm. In Activity One students use area models to visualize the multiplication of 2-digit numbers, and to understand the concept of partial products. In Activity Two students use an alternative algorithm related to the modeling in Activity One. In Activity Three students model the multiplication of 3-digit numbers by a 1-digit multiplier and use that modeling to understand regrouping and recording regrouping in multiplication. Activity Four uses a place-value frame to review the recording of each step in the multiplication algorithm. Use one or more of these activities with your students, as needed.

TEACHING THE *INTERVENTION WORKSHOP* LESSON

Getting Started

◆ **Say:** *What does 3 times 5 equal?* (15) *3 times 50?* (150) *30 times 50?* (1500) *30 times 500?* (15,000)

ACTIVITY ONE: Use Area Models to Find Partial Products

Materials: centimeter grid paper (BLM 1), several sheets per student; clear tape

Write the following exercise on the board.

$$\begin{array}{r} 34 \\ \times\ 23 \\ \hline \end{array}$$

Tell students that they can use a diagram to help them find the product. On grid paper have students draw a grid that is 34 squares wide and 23 squares high. (Students will have to tape together more than one sheet of grid paper in order to have sufficient space for a 34 × 23 grid.)

◆ **Say:** *How does the diagram represent the multiplication problem?* (Possible response: It shows 23 rows of 34 squares, or 34 rows of 23 squares.)

Tell students to draw a vertical dividing line that splits the width into two, 30 squares and 4 squares. Then tell students to draw a horizontal dividing line that splits the height into two, 20 squares and 3 squares. These lines separate the rectangle into four sections. Have students label the sections as indicated below:

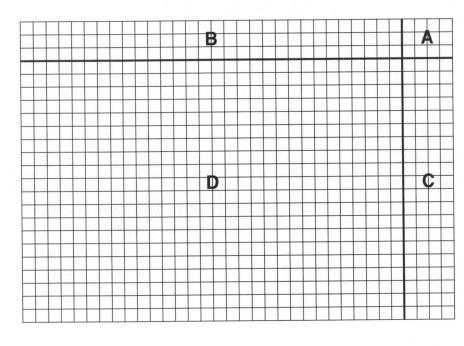

Section A:
3 squares high by
4 squares wide

Section B:
3 squares high by
30 squares wide

Section C:
20 squares high by
4 squares wide

Section D:
20 squares high by
30 squares wide

Have students find the number of squares in each section. Then ask students the questions below. As students give their responses, record the partial products on the board.

◆ **Say:** *How many squares are in Section A?* (12) *How did you find that number?* (possible responses: multiplied 3 × 4; counted squares) *How many squares are in Section B?* (90) *How did you find that number?* (possible responses: multiplied 3 × 30; counted squares) *How many squares are in Section C?* (80) *How did you find that number?* (possible responses: multiplied 20 × 4; counted squares) *How many squares are in Section D?* (600)

How did you find that number? (possible responses: multiplied 20 × 30; counted squares) *How can you find the total number of squares in the diagram?* (Add the squares in each part of the diagram.) *What is 23 times 34?* (782)

```
      34
    × 23
      12 ← 3 × 4     (A)
      90 ← 3 × 30    (B)
      80 ← 20 × 4    (C)
     600 ← 20 × 30   (D)
     782
```

Have students draw diagrams to find the following products: 17 × 25; 41 × 32.

ACTIVITY TWO: Alternative Algorithm: Use Partial Products to Multiply

Write the following on the board as students do so on sheets of paper.

```
    234
  ×  56
```

Have students multiply the ones.

◆ **Say:** *What is 6 times 4 ones?* (24 ones) *What number can you write to show 24 ones?* (24) *Write the number.*

◆ **Say:** *What is 6 times 3 tens?* (18 tens) *What number can you write to show 18 tens?* (180) *Write the number.*

◆ **Say:** *What is 6 times 2 hundreds?* (12 hundreds) *What number can you write to show 12 hundreds?* (1200) *Write it.*

```
    234
  ×  56
      24 ← 6 × 4 ones, or 6 × 4
     180 ← 6 × 3 tens, or 6 × 30
    1200 ← 6 × 2 hundreds, or 6 × 200
```

Have students multiply by the tens.

◆ **Say:** *What is 50 times 4 ones?* (200) *Write it. What is 50 times 3 tens?* (150 tens) *What number can you write to show 150 tens?* (1500) *Write the number. What is 50 times 2 hundreds?* (100 hundreds) *What number can you write to show 100 hundreds?* (10,000) *Write this number.*

```
    234
  ×  56
      24 ← 6 × 4 ones, or 6 × 4
     180 ← 6 × 3 tens, or 6 × 30
    1200 ← 6 × 2 hundreds, or 6 × 200
     200 ← 50 × 4 ones, or 50 × 4
    1500 ← 50 × 3 tens, or 50 × 30
   10000 ← 50 × 2 hundreds, or 50 × 200
```

Have students add the individual products to find the total product for 56 × 234. (13,104) Have students use this method independently to find other products involving 2-digit multipliers such as 23 × 36 and 19 × 48.

ACTIVITY THREE: Use Models to Show and Record Regrouping in Multiplication

Materials: base ten blocks

Write the following on the board.

h	t	o
1	5	8
		3

(× in front, 3 in ones column, with line below)

◆ **Say:** *How many sets of 158 would you need to show to find the answer to three times 158?* (3)

Have students use base ten blocks to model the problem.

◆ **Say:** *How many ones are there?* (24) *How many tens?* (15) *How many hundreds?* (3) *Regroup the ones as tens. How many tens can you make from the ones?* (2) *Now that you have regrouped, how many ones do you have?* (4)

In the problem on the board, write a 4 at the bottom of the ones column and a 2 at the top of the tens column.

◆ **Say:** *Combine the 2 tens you just made with the 15 you already had. How many tens do you have now?* (17) *Regroup tens as hundreds. How many hundreds can you make from the tens?* (1) *Now that you have regrouped, how many tens do you have?* (7)

In the problem on the board, write a 7 at the bottom of the tens column and a 1 at the top of the hundreds column.

◆ **Say:** *Combine the hundred you just made with the 3 hundreds you already had. How many hundreds do you have now?* (4)

In the problem on the board, write a 4 at the bottom of the hundreds column.

◆ **Say:** *What does 3 times 158 equal?* (474)

Repeat the activity to give examples of the following:

- multiplying a 3-digit number by a 1-digit number without regrouping, such as 4 × 121;

- multiplying a 3-digit number by a 1-digit number with regrouping of ones only, such as 4 × 216;

- multiplying a 3-digit number by a 1-digit number with regrouping of tens only, such as 2 × 384;

- multiplying a 3-digit number by a 1-digit number with regrouping of tens and ones, such as 5 × 196.

ACTIVITY FOUR: Use a Multiplication Frame to Record Regrouping

Materials: multiplication frames for thousands, hundreds, tens, and ones (BLM 3)

Have students write the multiplication exercise 24 × 148 in a multiplication frame, leaving room above the 148 for two rows of regrouped numbers.

Guide students through the multiplication and recording process, place by place. First multiply by the ones.

♦ **Say:** *What is 4 times 8?* (32) *To record this partial product, write 2 at the bottom of the ones place and 3 above the 4 in the tens place. Why did we write 2 at the bottom of the ones place?* (There are 2 ones in 32.) *Why did we write 3 at the top of the tens place?* (There are 3 tens in 32, and you regroup those 3 tens.)

th	h	t	o
	1	3	
	1	4	8
×		2	4
	5	9	2

Next have students multiply 4 × 40 and add the regrouped number.

♦ **Say:** *How many tens is 4 times 4 tens?* (16 tens) *Add the 3 regrouped tens to the 16 tens. How many tens is that in all?* (19 tens) *Regroup 10 tens as 1 hundred. How many tens do you have now?* (9) *Write that 9 at the bottom of the tens column, and the regrouped hundred at the top of the hundreds place.*

Now have students multiply 4 × 100 and add the regrouped number.

♦ **Say:** *How many hundreds is 4 times 100?* (4 hundreds) *Add the regrouped hundred to the 4 hundreds. How many hundreds is that in all?* (5) *Write that 5 at the bottom of the hundreds column.*

Now guide students through multiplying by the tens. Start by having students cross out the regrouped 1 and 3 at the top of the hundreds and tens column. Explain that they are doing this because they already used those regrouped numbers to make the partial product 592.

th	h	t	o
		1	
	1̶	3̶	
	1	4	8
×		2	4
	5	9	2
2	9	6	0

Now have students multiply 20 × 8.

♦ **Say:** *How many tens is 2 tens times 8?* (16 tens) *To record this partial product, write 60 (or 6 tens out of the 16 tens) at the bottom in the tens and ones places, and write 1 at the top of the tens place, above the crossed out 3. This 1 stands for 10 of the 16 tens, or 100.*

Now have students multiply 20 × 40 and add the regrouped number.

♦ **Say:** *What is 2 times 4?* (8) *What regrouped number do you add to 8?* (1) *What is 8 plus 1?* (9) *Write 9 at the bottom of the next column, the hundreds column. Why do you write this number at the bottom of the hundreds column?* (You are multiplying 20 times 40 to get 800, then adding 1 regrouped hundred to get 900.)

Now have students multiply 20 × 100.

♦ **Say:** *What is 2 times 1?* (2) *Write 2 at the bottom of the next column, the thousands column. Why do you write this number at the bottom of the thousands column?* (You are multiplying 20 times 100 to get 2,000.) *Add the partial products. What does 24 times 148 equal?* (3552)

SUMMING-IT-UP QUIZ

Give this quiz to students.

Write this exercise on the board.

$$
\begin{array}{r}
264 \\
\times\ 38 \\
\hline
2112 \\
7920 \\
\hline
\end{array}
$$

◆ **Say:** *How do you find the partial product 2112?* (Multiply 8×264.) *How do you find the partial product 7920?* (Multiply 30×264.) *How do you use the partial products 2112 and 7920 to find the product for 38 times 264?* (Add the partial products.) *What does 38 times 264 equal?* (10,032)

Write this exercise on the board.

$$
\begin{array}{r}
456 \\
\times\ 85 \\
\hline
\end{array}
$$

◆ **Say:** *Find the product for 5 times 456. What is the product?* (2280) *Find the product for 80 times 456. What is the product?* (36,480) *What does 85 times 456 equal?* (38,760)

If students answer the *Summing-It-Up Quiz* correctly, you may want to assign the review lessons on pages 11 and 12 of *Skills Update* in the student textbook. Then students should return to Lesson 2-3 and resume work on page 69 of their textbook. If students do not successfully complete the *Summing-It-Up Quiz,* further remediation may be necessary.

Use with Lesson 2-6, exercises 1–12, text pages 74–75.

DIAGNOSTIC INTERVIEW

Before beginning this *Intervention Workshop* lesson, have students complete the following exercises.

Diagnostic Exercises

Write this exercise on the board.

$$5\overline{)38}$$

Have students use counters to find the quotient and remainder. Then question students about the division process.

◆ **Say:** *How did you use counters to find the quotient?* (Possible responses: divided the counters into 5 equal groups; separated the counters into groups of 5) *What is the quotient? the remainder?* (The quotient is 7; the remainder is 3.)

Write this exercise on the board.

$$3\overline{)234}$$

◆ **Say:** *How can you decide where to begin the quotient?* (possible response: Look at the greatest place, which is the hundreds. See if there are enough hundreds to divide. Since 3 is greater than 2, there are not enough hundreds to divide, and you need to look at the tens. Since 3 is less than 23, there are enough tens to divide.)

◆ **Say:** *How do you find the first digit in the quotient?* (Find about how many threes there are in 23; there are about 7 threes.)

◆ **Say:** *What do you do next?* (Write 7 in the tens place in the quotient. Multiply $7 \times 3 = 21$. Subtract 21 from 23.)

◆ **Say:** *Complete the division by dividing the ones. What is the quotient?* (78)

If students answer the Diagnostic Exercises correctly, there is probably no need for them to do this *Intervention Workshop* lesson. Have those students rejoin their class on page 75 of the student textbook. For students who have had difficulties answering the exercises above, continue the *Intervention Workshop*.

BACKGROUND

Exercises 1–12 on page 75 of the student textbook require students to use short division to divide 2-, 3-, and 4-digit numbers by 1-digit divisors. Since short division is derived from the algorithm for division, these *Intervention*

Workshop activities address the needs of students who may not understand the meaning of division, may not have mastered the division algorithm, or do not understand the concept of remainders. Activity One reviews the concept of remainders, and provides opportunities to find 1-digit quotients with and without remainders. In Activity Two students use models to divide 2- and 3-digit numbers and find quotients with and without remainders. Activity Three provides an alternative algorithm for dividing. In Activity Four students practice recording division in which a 4-digit number is divided by a 1-digit divisor. Use one or more of these activities with your students, as needed.

TEACHING THE *INTERVENTION WORKSHOP* LESSON

Getting Started

◆ **Say:** *About how many fours are in 13?* (3) *About how many fives are in 32?* (6) *About how many threes are in 10?* (3)

ACTIVITY ONE: Use Models to Understand Division with and without Remainders

Materials: counters

Present the following problem.

- *José has 26 quarters. He has a coin album that fits 6 quarters per page. How many pages will he fill? How many quarters will be left over?*

Have students use counters to find the answer to the problem.

◆ **Say:** *How many counters would you use to show the total number of quarters José has?* (26 counters)

Have students show 26 counters. Remind them that each page holds 6 quarters. Have students separate the counters into groups of 6.

◆ **Say:** *How many groups of 6 did you make?* (4) *How many pages will José fill?* (4) *How many counters were left over?* (2) *How many quarters will be left over?* (2)

Write the following on the board.

26 ÷ 6 = 4 R2

◆ **Say:** *How does this division sentence represent the problem you just solved?* (The division sentence shows that when 26 is separated into groups of 6, you get 4 groups of 6 with 2 left over, or a remainder of 2.)

Remind students that R2 means a remainder of 2. Explain that if there are no leftovers, there is no remainder.

Have students solve the problems below. After students model each problem, have them write a division sentence to show what they did.

- *You have 31 counters. Find how many groups of 5 you can make. If there is a remainder, list it.* (31 ÷ 5 = 6 R1)

- *You have 24 counters. Find how many groups of 8 you can make. If there is a remainder, list it.* (24 ÷ 8 = 3)

- *You have 39 counters. Find how many groups of 9 you can make. If there is a remainder, list it.* (39 ÷ 9 = 4 R3)

ACTIVITY TWO: Use Models to Divide 2- and 3-Digit Numbers, with and without Remainders

Materials: base ten blocks, sheets of paper

Write the following on the board.

$$4\overline{)92}$$

Help students read the division exercise.

◆ **Say:** *In this division exercise, what is the total number that you start with?* (92)

Remind students that that number is called the *dividend*. Have students model the dividend by showing 9 ten rods and 2 ones units. Then have students use the models to divide.

◆ **Say:** *Into how many groups do you need to separate 92 in order to complete the division?* (4) *Do you have enough tens to share among four groups with at least 1 ten in each group?* (Yes)

Have students place ten rods one at a time on each of four pieces of paper, continuing until they cannot share the rods equally.

◆ **Say:** *How many tens are in each group?* (2) *How many tens are left over?* (1)

Now have students complete the division.

◆ **Say:** *In order to share the leftover ten, you must regroup the leftover ten as ten ones. Regroup the ten as 10 ones. How many ones do you have now?* (12)

Have students share the ones equally among the four groups.

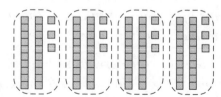

◆ **Say:** *How many tens are in each group?* (2) *How many ones?* (3) *How many ones*

are left over? (0) *Is there a remainder?* (No)

Repeat the activity to give the following:

- examples in which a 2-digit number is divided by a 1-digit number to get a 2-digit quotient without a remainder, such as 69 ÷ 3 and 68 ÷ 4;

- examples in which a 2-digit number is divided by a 1-digit number to get a 2-digit quotient with a remainder, such as 86 ÷ 7 and 53 ÷ 2.

Now have students use models to divide a 3-digit number by a 1-digit number.

Write the following on the board.

$$3\overline{)746}$$

Have students use the models to divide.

◆ **Say:** *What number do you need to divide?* (746) *Use models to show 746. Into how many groups do you need to separate 746 in order to complete the division?* (3) *Do you have enough hundreds to share among three groups with at least 1 hundred in each group?* (Yes)

Have students place hundred flats one at a time on each of three pieces of paper, continuing until they cannot share the flats equally.

◆ **Say:** *How many hundreds are in each group?* (2) *How many hundreds are left over?* (1) *In order to share the leftover hundred, you must regroup the hundred as 10 tens. Regroup the hundred as 10 tens. How many tens do you have now?* (14) *Do you have enough tens to share among 3 groups with at least 1 ten in each group?* (Yes)

Have students place ten rods one at a time on each piece of paper, continuing until they cannot share the rods equally.

◆ **Say:** *How many tens are in each group?* (4) *How many tens are left over?* (2) *In order to share the 2 tens that are left over, you must regroup each leftover ten as 10 ones. Regroup each of the 2 tens as 10 ones. How many ones do you have now?* (26)

Have students share the ones equally among the three groups.

◆ **Say:** *How many hundreds are in each group?* (2) *How many tens?* (4) *How many ones?* (8) *How many ones are left over?* (2)

What does 746 divided by 3 equal? (248, with a remainder of two)

Repeat the activity to give the following:

- examples in which a 3-digit number is divided by a 1-digit number to get a quotient without a remainder, such as $756 \div 2$ and $208 \div 8$;

- examples in which a 3-digit number is divided by a 1-digit number to get a quotient with a remainder, such as $956 \div 5$ and $614 \div 7$.

ACTIVITY THREE: Alternative Division Algorithm

Demonstrate the following division algorithm, which makes use of division "helping numbers" to estimate quotients in an alternative fashion.

◆ **Say:** *Using the special "helping numbers" 1, 2, 5, and 10 will help you find quotients when you are not sure how to estimate.*

Display the four helping numbers, the following division problem, and a three-column chart with the headings *hundreds*, *tens*, and *ones*.

$$8\overline{)2893}$$

◆ **Say:** *Which of the helping numbers do you think will show how many times 2893 can be divided by 8?*

Students' responses will depend upon the level of their number sense, but accept any one of the four helping numbers. For example, if students choose 2, demonstrate using 2 in the algorithm.

◆ **Say:** *Now multiply. What is the product of 8 and 2?* (16) *Can you subtract 16 from 2?* (No) *Can you subtract 16 from 28?* (Yes)

Have the student subtract and begin to record the process.

$$\begin{array}{r} 8\overline{)2893} \\ -16 \\ \hline 12 \end{array}$$

◆ **Say:** *Because the 8 is in the hundreds place of 2893, write the helping number you used, 2, in the hundreds column of the chart.*

hundreds	tens	ones
2		

◆ **Say:** *The difference, 12, is larger than 8, so pick another helping number to multiply by 8.* (1) *What is the product of 8 times 1?* (8) *Can you subtract 8 from 12?* (Yes) *What is the difference?* (4)

Continue to record the process.

$$\begin{array}{r} 8\overline{)2893} \\ -16 \\ \hline 12 \\ -8 \\ \hline 4 \end{array}$$

◆ **Say:** *Because you were still working under the 8 in the hundreds place in 2893, write the second helping number you used, 1, in the hundreds column of the chart, below the 2.*

hundreds	tens	ones
2		
1		

◆ **Say:** *Look again at the division problem. Can you divide 4 by 8?* (No) *What must you do to continue dividing?* (Bring down the next number in the dividend.) *Bring down the 9 to form the number 49. Since 49 is larger than 8, pick a helping number.* (Suggest 5, although either of the other helping numbers, 1 or 2, will also eventually lead to a correct quotient.)

◆ **Say:** *What is the product of 8 and 5?* (40) *Can you subtract 40 from 49?* (Yes) *What is the difference?* (9)

Continue to record the process.

$$8\overline{)2893}$$
$$\underline{-16}$$
$$12$$
$$\underline{-\ 8}$$
$$49$$
$$\underline{-40}$$
$$9$$

◆ **Say:** *Because we brought down 9, which is in the tens place in 2893, write the helping number you used, 5, in the tens column in the chart.*

hundreds	tens	ones
2	5	
1		

◆ **Say:** *The difference, 9, is larger than 8, so pick another helping number to multiply by 8.* (1) *What is the product of 8 times 1?* (8) *Can you subtract 8 from 9?* (Yes) *What is the difference?* (1) *Where will you record the helping number you used this time in the chart?* (in the tens column under the 5) *Why?* (We are still working under the tens place in the original dividend.)

hundreds	tens	ones
2	5	
1	1	

◆ **Say:** *Since the difference, 1, is less than 8, bring down the 3 from the ones place in 2893 so that you can continue dividing.*

Continue to record the process.

$$8\overline{)2893}$$
$$\underline{-16}$$
$$12$$
$$\underline{-\ 8}$$
$$49$$
$$\underline{-40}$$
$$9$$
$$\underline{-\ 8}$$
$$13$$

◆ **Say:** *Which helpful number will you use next?* (1) *Where will you record it in the chart? Why?* (in the ones place because the 3 that was brought down is in the ones place of the dividend) *What is the product of 8 times 1?* (8) *Subtract 8 from 13. What is the difference?* (5) *Can you continue dividing?* (No; 5 is less than 8 and there are no more numbers to bring down.)

Complete recording the process by writing 8 below 13 and showing the difference of 5.

◆ **Say:** *You can use the helpful numbers in the chart to find the quotient. Just add the numbers in each column and be sure to write the remainder, 5.*

hundreds	tens	ones	
2	5	1	
1	1		
3	6	1	R 5

Say: *So 2893 divided by 8 equals 361 R5.*

Have students do the same exercise using a different sequence of helpful numbers, or give them another division problem such as 3647 ÷ 7 or 5037 ÷ 9.

ACTIVITY FOUR: Use a Place-Value Frame to Record Division

Materials: place-value frame for thousands, hundreds, tens, and ones (BLM 4)

Write the following in a place-value frame on the board.

$$6\overline{)2951}$$

Have students copy the division example into their place-value frames. Then lead students through the process of dividing and recording. Have students look at the thousands place.

◆ **Say:** *Can you share 2 thousands equally among 6 groups with at least 1 thousand in each group?* (No) *Will the quotient begin in the thousands place?* (No) *How can you divide the thousands by 6?* (Regroup thousands as hundreds.)

Have students circle the digits in the thousands and hundreds places as you do so at the board.

th	h	t	o
	4	9	1 R5
6) ②⃝	⑨⃝	5	1
2	4		
	⑤⃝	⑤⃝	
	5	4	
		①⃝	①⃝
			6
			5

◆ **Say:** *The circled number is the number of hundreds in the dividend. How many hundreds are there?* (29) *Are there enough hundreds to divide?* (Yes) *Where will the quotient begin?* (in the hundreds place)

Have students divide the hundreds by 6.

◆ **Say:** *How many hundreds will be in each group if you share 29 hundreds equally among 6 groups?* (4)

Record the 4 in the quotient on the board as students do so in their place-value frames.

◆ **Say:** *How many hundreds are in 6 groups of 4 hundreds?* (24 hundreds)

Record the 24 on the board as students do so in their place-value frames.

◆ **Say:** *How many hundreds will be left over?* (5)

Record the 5 as students do so in their place-value frames.

Have students bring down the 5 tens, then circle the digits in the hundreds and tens places as you do so at the board. Explain that this is the number of tens left.

◆ **Say:** *How many tens are there?* (55) *Are there enough tens to divide?* (Yes) *Divide the tens by 5. How many tens will be in each group if you share 55 tens equally among 6 groups?* (9)

Record the 9 in the quotient as students do so in their place-value frames.

◆ **Say:** *How many tens are in 6 groups of 9 tens?* (54 tens)

Record the 54 on the board as students do so in their place-value frames.

◆ **Say:** *How many tens will be left over?* (1)

Record the 1 as students do so in their place-value frames. Have students bring down the one, then circle the digits in the tens and ones places as you do so at the board.

◆ **Say:** *This is the number of ones left. How many ones are there?* (11) *Are there enough ones to divide?* (Yes) *Divide the ones by 6. How many ones will be in each group if you share 11 ones equally among 6 groups?* (1)

Record the 1 in the quotient as students do so in their place-value frames.

◆ **Say:** *How many ones are in 6 groups with 1 one in each group?* (6 ones)

Record the 6 on the board as students do so in their place-value frames.

◆ **Say:** *How many ones will be left over?* (5)

Record the 5 as students do so in their place-value frames.

◆ **Say:** *What does 2951 divided by 6 equal?* (491, with a remainder of 5)

SUMMING-IT-UP QUIZ

Give this quiz to students.

Write the following on the board.

$$4 \overline{)3042}$$

◆ **Say:** *In what place will the quotient start? How do you know?* (the hundreds place; there are not enough thousands to divide) *What will be the first digit of the quotient? How do you know?* (7; Possible answers: There are about 7 fours in 30; $7 \times 4 = 28$ is the closest you can get to 30 without exceeding 30.)

Write the following in a place-value frame on the board.

$$6 \overline{)951}$$

Have students copy this division example in their place-value frames.

◆ **Say:** *To solve this problem, what do you do first?* (Possible response: Check the hundreds to see if there are enough hundreds to divide. Since there are enough, divide the hundreds.)

Have students complete the division example. (158 R3)

If students answer the *Summing-It-Up Quiz* correctly, they should then resume work on page 75 in their textbook. If students do not successfully complete the *Summing-It-Up Quiz*, further remediation may be necessary.

Multiplication and Division Fact Families

Use with Lesson 2-6, exercises 1–12, text pages 74–75.

DIAGNOSTIC INTERVIEW

Before beginning this *Intervention Workshop* lesson, have students complete the following exercises.

Diagnostic Exercises

Write this example on the board.

$$8 \times 6 = 48$$

◆ **Say:** *What is another multiplication fact you can write using the same three numbers?* ($6 \times 8 = 48$) *What division facts can you write using the same three numbers?* ($48 \div 6 = 8$; $48 \div 8 = 6$)

Write this exercise on the board.

$$5\overline{)21}$$

◆ **Say:** *About how many fives are in 20?* (about 4 fives) *What multiplication fact can help you complete this division exercise?* (possible answer: $4 \times 5 = 20$) *Complete the division. What is the answer?* (4 with a remainder of 1)

If students answer the Diagnostic Exercises correctly, there is probably no need for them to do this *Intervention Workshop* lesson. Have those students rejoin their class on page 75 of the student textbook. For students who have had difficulties answering the exercises above, continue the *Intervention Workshop*.

BACKGROUND

Exercises 1–12 on page 75 of the student textbook require students to use short division to divide 2-, 3-, and 4-digit numbers by 1-digit divisors. Understanding the relationship between multiplication and division and assimilating multiplication and division fact families are prerequisite skills for both short division and for using the division algorithm from which short division is derived. Students can use multiplication facts to help them complete the division in each place of the dividend. To address the needs of students who have not yet mastered the relationship between multiplication and division, the following *Intervention Workshop* activities have been provided. In Activity One students review the concept of multiplication and division fact families. In Activity Two students use the relationship between multiplication and division to help them find 1-digit quotients, with and without remainders. In Activity Three students use multiplication facts to help them find each digit in a 3-digit quotient. Use one or more of these activities with your students, as needed.

TEACHING THE *INTERVENTION WORKSHOP* LESSON

Getting Started

◆ **Say:** *About how many fours are in 13?* (about 3 fours) *About how many fives are in 32?* (about 6 fives) *About how many threes are in 10?* (about 3 threes)

ACTIVITY ONE: Use Models to Understand Multiplication and Division Fact Families

Materials: connecting cubes, 1-9 spinner

Have students make 4 cube trains with 3 cubes in each train.

◆ **Say:** *How many cube trains are there?* (4) *How many cubes are in each train?* (3) *What multiplication fact can you use to show the total number of cubes? Write that multiplication fact.* ($4 \times 3 = 12$)

Now have students use the cube trains to write a related division fact.

◆ **Say:** *What division fact can you write to show separating 12 cubes into groups of 3 cubes?* ($12 \div 3 = 4$) *Write that division fact. How are the multiplication fact and the division fact that you just wrote alike?* (They use the same three numbers.) *What is another multiplication fact that you can write that has the same three numbers?* ($3 \times 4 = 12$) *Write that fact. What is another division fact that you can write that has the same three numbers?* ($12 \div 4 = 3$)

Remind students that the four facts that they just wrote are called a *fact family*, and that a fact family is a group of facts that uses the same three (or two) numbers. Then have students work in pairs to find fact families. One student spins a spinner twice and writes two multiplication facts for those two factors, or one fact if the two factors are the same. The second student completes the fact family by writing two division facts. Partners switch roles and repeat the activity.

ACTIVITY TWO: Use Multiplication Facts to Find Quotients and Remainders

Write the following on the board.

$$4\overline{)19}$$

◆ **Say:** *This division example asks how many fours are in 19. To complete the division find a multiplication fact that has a 4 as a factor and a product that is close to, but not greater than, 19. What multiplication fact will you use to help* *you find the quotient?* ($4 \times 4 = 16$) *Why can you not use 4 times 5?* ($4 \times 5 = 20$, and 20 is greater than 19.) *What is the quotient?* (4) *How do you find the remainder?* (Subtract: $19 - 16 = 3$.)

Write the following on the board.

$$6\overline{)32}$$

◆ **Say:** *The division exercise asks how many sixes are in 32. What multiplication fact that has a factor of 6 will you use to help you find the quotient?* ($5 \times 6 = 30$) *Why can you not use 6 times 6?* ($6 \times 6 = 36$, and 36 is greater than 32.) *What is the quotient?* (5) *How do you find the remainder?* (Subtract: $32 - 30 = 2$.)

Repeat the activity with other examples involving 2-digit dividends, 1-digit divisors, and 1-digit quotients with and without remainders, such as $35 \div 8$, $42 \div 7$, $61 \div 9$, and $41 \div 6$. Have students tell what multiplication fact they would use to help them complete each division.

ACTIVITY THREE: Use Multiplication Facts to Understand the Division Algorithm

Materials: place-value frame for thousands, hundreds, tens, and ones (BLM 4)

Write the following in a place-value frame on the board.

$$7\overline{)3725}$$

Have students copy the division exercise into their place-value frames. Point out the thousands.

◆ **Say:** *Can you divide 3 thousands into 7 groups and get at least 1 thousand in each group?* (No) *Will the quotient begin in the thousands place?* (No) *How can you divide the thousands by 7?* (Regroup thousands as hundreds.)

Have students circle the digits in the thousands and hundreds places as you do so at the board.

th	h	t	o
	5	3	2 R1
7) 3	7	2	5
3	5		
	2	2	
	2	1	
		1	5
		1	4
			1

◆ **Say:** *The circled number is the number of hundreds in the dividend. How many hundreds are there?* (37) *What multiplication fact with a factor of 7 has a product close to, but not greater than, 37?* ($5 \times 7 = 35$) *5 times 7 equals 35. How many hundreds will there be in each group if you divide 37 hundreds by 7?* (5)

Record the 5 in the quotient on the board as students do so in their place-value frames.

◆ **Say:** *How many hundreds are in 7 groups of 5 hundreds?* (35 hundreds)

Record the 35 on the board as students do so in their place-value frames.

◆ **Say:** *How many hundreds will be left over?* (2)

Record the 2 as students do so in their place-value frames. Have students bring down the 2 tens, then circle the digits in the hundreds and tens places as you do so at the board. Explain that this is the number of tens left.

◆ **Say:** *How many tens are there?* (22) *What multiplication fact with a factor of 7 has a product close to, but not greater than, 22?* ($3 \times 7 \times 21$) *3 times 7 equals 21. How many tens will there be in each group if you divide 22 tens by 7?* (3)

Record the 3 in the quotient on the board as students do so in their place-value frames.

◆ **Say:** *How many tens are in 7 groups of 3 tens?* (21 tens)

Record the 21 on the board as students do so in their place-value frames.

◆ **Say:** *How many tens will be left over?* (1)

Record the 1 as students do so in their place-value frames. Have students bring down the 5, then circle the digits in the tens and ones places as you do so at the board.

◆ **Say:** *This is the number of ones left. How many ones are there?* (15) *What multiplication fact with a factor of 7 has a product close to, but not greater than, 15?* ($7 \times 2 = 14$) *7 times 2 equals 14. How many ones will there be in each group if you divide 15 ones by 7?* (2)

Record the 2 in the quotient as students do so in their place-value frames.

◆ **Say:** *How many ones are in 7 groups of 2 ones?* (14 ones)

Record the 14 on the board as students do so in their place-value frames.

◆ **Say:** *How many ones will be left over?* (1)

Record the 1 as students do so in their place-value frames.

◆ **Say:** *What does 3725 divided by 7 equal?* (532, with a remainder of 1)

SUMMING-IT-UP QUIZ

Do this quiz with students.

Write the following on the board: 6, 7, 42

◆ **Say:** *What multiplication and division facts can you write using these three numbers?* ($6 \times 7 = 42$, $7 \times 6 = 42$, $42 \div 7 = 6$; $42 \div 7 = 6$)

Write the following on the board.

$$9\overline{)37}$$

◆ **Say:** *What multiplication fact will you use to help you find the quotient?* ($4 \times 9 = 36$) *Why can you not use 5 times 9?* ($5 \times 9 = 45$, and 45 is greater than 37.) *What is the quotient?* (4) *How do you find the remainder?* (Subtract: $37 - 36$.)

Write the following in a place-value frame on the board.

$$6\overline{)527}$$

Have students copy this division example in their place-value frames.

◆ **Say:** *Can you divide 5 hundreds into 6 groups and get at least 1 hundred in each group?* (No) *Will the quotient begin in the hundreds place?* (No) *How can you divide the hundreds?* (Regroup hundreds as tens.) *There are 52 tens. What multiplication fact can help you find 52 divided by 6?* ($8 \times 6 = 48$) *What will be the first digit in the quotient?* (8)

If students answer the *Summing-It-Up Quiz* correctly, they should then resume work on page 75 in their textbook. If students do not successfully complete the *Summing-It-Up Quiz*, further remediation may be necessary.

Estimating with Compatible Numbers

Use with Lesson 2-7, exercises 1–12, text pages 76–77.

DIAGNOSTIC INTERVIEW

Before beginning this *Intervention Workshop* lesson, have students complete the following exercises.

Diagnostic Exercises

◆ **Say:** *Tell which pair of numbers is compatible and why. If the pair is not compatible, name a pair of compatible numbers near the pair.*

　　　　35 and 6　　　　　　*72 and 8*

(35 and 6 are not a compatible pair of numbers because 6 does not divide 35 evenly with no remainder. A compatible pair near 35 and 6 is 36 and 6; 72 and 8 are a compatible pair because 8 divides 72 evenly with no remainder.)

◆ **Say:** *Use compatible numbers to estimate the quotient. Remember: A* quotient *is the answer to a division exercise.*

　　　　252 ÷ 6　　　　　　*5238 ÷ 73*

(252 ÷ 6 → 240 ÷ 6 = 40;
5238 ÷ 73 → 4900 ÷ 70 = 70 or 5600 ÷ 70 = 80)

If students answer the Diagnostic Exercises correctly, there is probably no need for them to do this *Intervention Workshop* lesson. But before they resume work on page 77, you may want to assign the review lessons on pages 13 and 14 of *Skills Update* in the student textbook. Then have students rejoin their class on page 77. For students who have had difficulty, continue this *Intervention Workshop*.

BACKGROUND

This *Intervention Workshop* lesson addresses the problem of how to choose a pair of compatible numbers for estimating the quotient of two whole numbers. Students begin by using arrays of objects to model a simple division (a 2-digit number by a 1-digit number) by forming as many equal groups as possible. If the pair of numbers is not compatible, students are guided to find a pair of compatible numbers near the given pair. Then students use patterns to find the quotients for sets of division problems whose dividends and divisors are multiples of powers of 10. At the end of the lesson, students use the strategies they have been working with to estimate whole-number quotients. They are guided to first find the quotient of a given simpler division and then use patterns of powers of 10 to find the estimated quotient of the original exercise.

TEACHING THE *INTERVENTION WORKSHOP* LESSON

Getting Started

Materials: counters, centimeter grid paper (BLM 1)

◆ **Say:** *Powers of 10 are 10, 100, 1000, and so on — the values of the places in our place-value system. Knowing how to multiply and divide whole numbers by powers of 10 will help you estimate quotients. Multiply 41 by 10, 100, and 1000.* (410; 4100; 41,000) *Divide 56,000 by 10, 100, and 1000.* (5600; 560; 56)

◆ **Say:** *A compatible pair of numbers is made up of two numbers that you can divide evenly with no remainder. Some division facts are examples of compatible numbers. Write a pair of compatible numbers that is a division fact and tell why the pair is compatible.* (sample answer: 48 and 6 because 6 divides 48 evenly with no remainder) *Use the division fact you wrote to find two more pairs of compatible numbers that are multiples of powers of 10.* (sample answer: 480 and 6; 4800 and 60)

⏩ USING PAGES IW3 and IW4

Have a student read aloud the sentence in the box at the top of page IW3. Relate the definition given to the division fact students wrote earlier in *Getting Started* in the following way.

◆ **Say:** *Tell the quotient of the compatible pair of numbers and of the powers of 10 multiples that you wrote.* (sample answer: 48 ÷ 6 = 8; 480 ÷ 6 = 80; 4800 ÷ 60 = 80)

Exercises 1–2 on page IW3 and Exercise 3 on page IW4
Have a student read aloud the direction line above exercise 1. Exercises 1–3, using arrays of squares to model, lead students to analyze whether the given pair of numbers, a division fact, or near fact, is compatible. If the pair is not compatible, the exercises guide students to name a nearby compatible pair of numbers. A follow-up in each exercise has students find a pattern of multiples of powers of 10 for the pair of compatible numbers so that they can work with the division of numbers with a greater number of digits.

◆ **Say:** *You may want to use counters to show the numbers given. This can help you see the number of equal groups and how many are left over.*

Assist students as necessary to complete exercises 1–3.

◆ **Say:** *Exercise 3 is just like exercises 1–2, but now you complete the drawing of the squares to show equal groups.*

After students complete exercise 3, ask the following.

◆ **Say:** *Is there another pair of compatible numbers nearby? If so, what is that pair?* (Yes, 42 and 6 or 48 and 6.)

Be sure that students do not use a mix of both pairs of compatible numbers when they complete the family of compatible pairs in exercise 3.

For students who have difficulty with exercise 3, use the following additional example for them to model with counters or shade in squares on grid paper.

34 and 4

_____ ÷ 4 = _____

_____0 ÷ 4 = _____

_____00 ÷ 4 = _____

_____,000 ÷ 4 = _____

Exercises 4–7 on page IW4

Have a student read aloud the direction line above exercises 4-7. If necessary remind students that a *quotient* is the answer to a division.

◆ **Say:** *You will start each exercise by completing a compatible number pair. Then you will complete a simpler division using the compatible number pair. Finding the quotient or answer to this simpler division will help you estimate the quotient of the original exercise. You will use patterns of powers of 10 just as you did in the earlier exercises.*

Guide students as necessary to complete exercises 4-7.

SUMMING-IT-UP QUIZ

Give this quiz to students.

◆ **Say:** *Estimate these quotients using compatible numbers. Tell the compatible pair of numbers you used for each estimated quotient.*

     a. 23 ÷ 5   *b. 671 ÷ 8*   *c. 4356 ÷ 58.*

(some possible answers: a: 20 ÷ 5 = 4; b: 640 ÷ 8 = 80; c: 4200 ÷ 60 = 70)

◆ **Say:** *Look at the three divisions you wrote. Tell which is a simple division fact and which is a power of 10 of a division fact.* (*a* is a simple division fact, 20 ÷ 5; *b* and *c* are each a power of 10 of a division fact: 64 ÷ 8 and 42 ÷ 6.)

If students answer the *Summing-It-Up Quiz* correctly, you may want to assign the review lessons on pages 13 and 14 of *Skills Update* in the student textbook. Then students should return to Lesson 2-7 and resume work on page 77 of their textbook. If students do not successfully complete the *Summing-It-Up Quiz*, further remediation may be necessary.

Answers: Pages IW3 and IW4

1. 1. 4; 4
    4; 40; 400

2. 56
    56; 56; 7
    56; 56; 70
    56; 56; 70
    56; 56; 700

3. cannot; 42 or 48
    42; 42; 7 or 48; 48; 8
    42; 42; 70; or 48; 48; 80
    42; 42; 70; or 48; 48; 80
    42; 42; 700; or 48; 48; 800

4. 40; 40; 5
    40; 5
    40; 5

5. 81; 81; 9
    81; 9
    81; 9

6. 300; 300; 6
    300; 6
    300; 6

7. 800; 800; 20
    800; 20
    800; 20

Understanding Decimal Place Value

Use with Lesson 3-1, exercises 1–11, text pages 94–95.

DIAGNOSTIC INTERVIEW

Before beginning this *Intervention Workshop* lesson, have students complete the following exercises.

Diagnostic Exercises

◆ **Say:** *Write each fraction or mixed number as a decimal.*

    9/100   6 7/1000

(0.09; 6.007)

◆ **Say:** *Write these numbers. Name the place and value of the digit 4 in each number.*

    0.45   0.054   5.045

(From left: 4 is in the tenths place, the value is 0.4; 4 is in the thousandths place, the value is 0.004; 4 is in the hundredths place, the value is 0.04.)

If students answer the Diagnostic Exercises correctly, there is probably no need for them to do this *Intervention Workshop* lesson. But before they resume work on page 95, you may want to assign the review lesson on page 6 of *Skills Update* in the student textbook. Then have students rejoin their class on page 95. For students who have had difficulty, continue this *Intervention Workshop* lesson.

BACKGROUND

This *Intervention Workshop* lesson addresses the problem of the relationship between place and value in decimals. Students begin by writing fractions and mixed numbers with denominators of powers of 10 as decimals by using place-value charts. Next students determine the place and value of a given digit by analyzing the digits of a decimal as the decimal appears in a place-value chart.

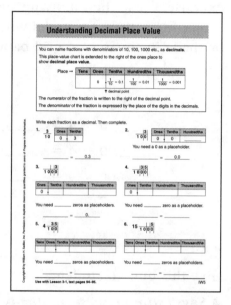

TEACHING THE *INTERVENTION WORKSHOP* LESSON

Getting Started

Materials: centimeter grid paper (BLM 1), place-value chart (BLM 5)

◆ **Say:** *The* numerator *of a fraction is the number above the fraction bar, and the* denominator *is the number below the fraction bar. All proper fractions are between zero and 1 on the number line.*

◆ **Say:** *Write any fraction with a denominator of 10.* (sample answer: 6/10) *Write any fraction with a denominator of 100.* (sample answer: 9/100) *Write any fraction with a denominator of 1000.* (sample answer: 68/1000)

🕐 USING PAGES IW5 and IW6

Have a student read aloud the material in the box at the top of page IW5. Relate the information that is given to the fractions students wrote earlier in *Getting Started* in the following way.

◆ **Say:** *A* decimal *is a number with a decimal point separating the ones place from the tenths place. If you name the fractions you wrote earlier as decimals, tell what digits would be written to the right of the decimal point.* (sample answer: 6; 9; 68)

Exercises 1–6 on page IW5

Have a student read aloud the direction line above exercises 1-6. These six exercises lead students to name fractions (exercises 1-4) and mixed numbers (exercises 5-6) with denominators of powers of 10 as decimals. Students use place-value charts. Vertical line segments align the digits of the numerator with the digits of the denominator to aid students in counting and then writing the correct number of zeros as placeholders in the decimal.

Before beginning exercise 4, you may want to provide an interim exercise with two digits in the numerator (interim exercise: 35/100).

For students who have difficulty with exercises 5-6, use these additional examples: for exercise 5, 4 5/10; for exercise 6, 35 3/100.

◆ **Say:** *Before you name the fraction or mixed number as a decimal in a place-value chart, you may want to write the fraction or mixed number on grid paper to help you align the digits in the numerator with the digits in the denominator. This will help you see how many zeros, if any, you need as placeholders.*

Exercises 7–15 on page IW6

Have a student read aloud the material in the box at the top of page IW6.

◆ **Say:** Place *means where the digit is positioned in a decimal. When a digit is assigned to a place in a decimal, it takes on a new value. For example in the decimal 0.569, the 5 is in the tenths place. The value of 5 in the decimal 0.569 is* 5×0.1 *or 0.5.*

You can explain the place and value of the digits 6 and 9 the same way.

Have students complete exercises 7–10 in which the decimals are already written in place-value charts. Assist students as necessary with exercises 11–15 in which they write the decimals in place-value charts and then determine the place and value of a given digit.

SUMMING-IT-UP QUIZ

Give this quiz to students.

◆ **Say:** *Write these three decimals. Name the place and value of the digit 6 in each decimal.*

     9.0861   *0.0106*   *5.06001*

(From left: 6 is in the thousandths place, the value is 0.006; 6 is in the ten thousandths place, the value is 0.0006; 6 is in the hundredths place, the value is 0.06.)

◆ **Say:** *Look at the three decimals you wrote. Name the decimal and the digit that has a value of 0.0001.* (The decimal is 9.0861, and the digit is 1.)

If students answer the *Summing-It-Up Quiz* correctly, you may want to assign the review lesson on page 6 of *Skills Update* in the student textbook. Then students should return to Lesson 3-1 and resume work on page 95 of their textbook. If students do not successfully complete the *Summing-It-Up Quiz,* further remediation may be necessary.

Answers: Pages IW5 and IW6

1. 3/10

2. 3; 3/100; 3

3. 0 0 3;
2; 3/1000; 003

4. 0 3 5;
1; 35/1000; 0.035

5. 4 3 5;
no; 4 35/100; 4.35

6. 1 5 0 0 5;
2; 15 5/1000; 15.005

7. 0.7

8. ones; 7

9. thousandths; 0.007

10. hundredths; 0.07

11. 0 0 3 0 7;
ten-thousandths; 0.0007

12. 3 0 0 7;
hundredths; 0.07

13. 0 3 0 7;
thousandths; 0.007

14. 0 0 0 3 0 7;
hundred-thousandths;
0.00007

15. 3 0 3 7 0 3;
thousandths; 0.007

Comparing and Ordering Decimals

Use with Lesson 3-4, exercises 1–19, text pages 100–101.

DIAGNOSTIC INTERVIEW

Before beginning this *Intervention Workshop* lesson, have students complete the following exercises.

Diagnostic Exercises

◆ **Say:** *Write these pairs of decimals. Compare each pair of decimals by writing* <, =, *or* >.

     0.5 and 0.2   0.1 and 0.19   5.7 and 5.07
(0.5 > 0.2; 0.1 < 0.19; 5.7 > 5.07)

◆ **Say:** *Write these decimals in order, from least to greatest: 4.52, 4.05, and 4.025.* (4.025, 4.05, 4.52)

If students answer the Diagnostic Exercises correctly, there is probably no need for them to do this *Intervention Workshop* lesson. Have those students rejoin their class on page 101 of the student textbook. For students who have had difficulty answering the exercises above, continue the *Intervention Workshop*.

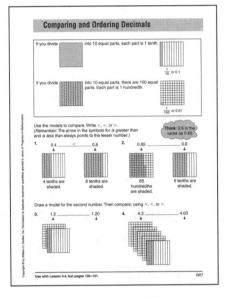

BACKGROUND

This *Intervention Workshop* lesson addresses the problem of how the place and value of the digits in decimals determine their relative value and order. Students begin by comparing decimals as shown on 10 × 10 grids. Comparing the sizes of the shaded areas on the base ten grids provides a clear visual methodology for students to readily determine which decimal is greater. Next students develop their understanding of a decimal number line by using patterns to label tick marks. At the end of the lesson students complete decimal patterns in preparation for using a number line to order decimals.

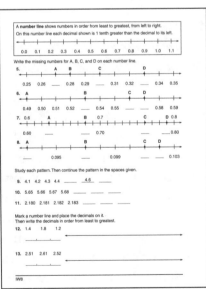

TEACHING THE *INTERVENTION WORKSHOP* LESSON

Getting Started

Materials: centimeter grid paper (BLM 1)

◆ **Say:** *On grid paper outline two 10 by 10 squares. Label one square A and the other square B. Shade as many small squares inside A and B as you wish. Tell which shaded area is larger and how you know that.* (Sample answer: The shaded area of A is larger because it has 27 shaded squares and B has only 5 shaded squares.)

Draw three number lines, each with three tick marks. Label the middle tick marks 100, 150, and 1001, respectively.

◆ **Say:** *A number line shows numbers in order, from least to greatest. Name the whole number just before and just after the numbers shown on these number lines.* (99 and 101; 149 and 151; 1000 and 1002)

▶ USING PAGES IW7 and IW8

Have a student read aloud the material in the box at the top of page IW7. Relate the models shown to the 10 × 10 grids students drew in *Getting Started* in the following way.

◆ **Say:** *If you let each small square inside A and B represent 1/100, say aloud the decimals that A and B show.* (sample answer: 27 hundredths, 5 hundredths)

Exercises 1–4 on page IW7

Have a student read aloud the direction line above exercises 1-2 and 3-4. By using base ten grid models and comparing the sizes of the shaded areas, students are guided to determine which decimal is greater. Before beginning these exercises you may want to review the meaning of the symbol for *is greater than* (>) and *is less than* (<) by writing, for example, 0.8 < 1.3 and 1.3 > 0.8 to demonstrate.

◆ **Say:** *When you read 0.8 is less than 1.3, you say* 8 tenths is less than 1 and 3 tenths. *When you read 1.3 is greater than 0.8, you say* 1 and 3 tenths is greater than 8 tenths.

In exercises 1-2 models for both decimals are provided, while in exercises 3-4 students draw the model for the second decimal before they make the comparison.

Before beginning exercise 2, you may want to provide an interim exercise to show equal decimals (interim exercise: 0.50 _____ 0.5).

◆ **Say:** *Because the shaded areas are the same size, the decimals are equal. Now you know that when you write one or more zeros to the right of the last digit of a decimal, the two decimals are equal. For example, 0.5 = 0.500.*

Assist students as necessary to complete exercises 1-4.

Exercises 5–13 on page IW8

Have a student read aloud the material in the box at the top of page IW8. Then have a student read aloud the direction line above exercises 5-8.

◆ **Say:** *Remember the whole numbers you wrote earlier in the* Getting Started *activity. Use that same idea to write the missing numbers on the number lines in these exercises.*

After students complete exercise 7, say the following.

◆ **Say:** *Each decimal written above the number line has another decimal written just below it on the number line. What can you say about these pairs of decimals?* (In exercise 7, the pairs of decimals are equal; 0.6 = 0.60 and 0.7 = 0.70.)

Have a student read aloud the direction line above exercises 9–11. Assist students as necessary to complete these exercises. Finally have a student read the direction line above exercises 12–13.

◆ **Say:** *You may want to use grid paper to model each of the three decimals. Then you can compare the sizes of the shaded areas, and this will tell you how to write the decimals in order, from least to greatest.*

Guide students as necessary to complete exercises 12–13.

SUMMING-IT-UP QUIZ

Give this quiz to students.

◆ **Say:** *Write these decimals and compare them by writing* <, =, *or* >.
  a. 3.6 and 3.60 b. 0.36 and 0.3 c. 0.036 and 0.063
(a: 3.6 = 3.60; b: 0.36 > 0.3; c: 0.036 < 0.063)

◆ **Say:** *Which pair(s) of decimals* a, b, *and/or* c *has other decimals between them?* (b and c)

◆ **Say:** *Write these decimals in order, from least to greatest:*
2.3, 0.23, and 2.03 (0.23, 2.03, 2.3)

If students answer the *Summing-It-Up Quiz* correctly, they should return to Lesson 3-4 and resume work on page 101 of their textbook. If students do not successfully complete the *Summing-It-Up Quiz,* further remediation may be necessary.

Answers: Pages IW7 and IW8

2. >   **3.** =   **4.** >   **5.** 0.27; 0.30; 0.33

6. 0.53; 0.56; 0.57 **7.** 0.63; 0.68; 0.76; 0.79 **8.** 0.093; 0.097; 0.101; 0.102

9. 4.5; 4.7 **10.** 5.69; 5.70; 5.71 **11.** 2.184; 2.185; 2.186

12. 1.2; 1.4; 1.8 **13.** 2.51; 2.52; 2.61

Decimals Multiplied by Whole Numbers

Use with Lesson 4-3, exercises 6–23, text pages 126–127.

DIAGNOSTIC INTERVIEW

Before beginning this *Intervention Workshop* lesson, have students complete the following exercises.

Diagnostic Exercises

Write this exercise on the board.

$$0.6 \times 50$$

◆ **Say:** *Will this product be greater than 50 or less than 50?* (less than 50) *How do you know?* (Possible response: $1 \times 50 = 50$. Since 0.6 is less than 1, 0.6×50 will have a product that is less than 50.) *Will this product be greater than 25 or less than 25?* (greater than 25) *How do you know?* (Possible response: 25 is 1/2 of 50; 0.6 is greater than 1/2.)

Write this exercise on the board.

$$\begin{array}{r} 6.974 \\ \times \quad 21 \\ \hline ? \end{array}$$

◆ **Say:** *Is 14,645.4 a reasonable value for this product?* (No) *How do you know?* (Possible response: 21×6.974 is close to 20×7 and $20 \times 7 = 140$; 140 is not close to 14,645.4.)

If students answer the Diagnostic Exercises correctly, there is probably no need for them to do this *Intervention Workshop* lesson. Have those students rejoin their class on page 127 of the student textbook. For students who have had difficulties answering the exercises above, continue the *Intervention Workshop*.

BACKGROUND

Exercises 6–23 on page 127 of the student textbook require students to multiply decimals by whole numbers. These *Intervention Workshop* activities address the needs of students who still have not mastered using number sense and estimation to predict the magnitude of decimal products, placing a decimal point in a product, and checking the reasonableness of answers. In Activity One students use number sense to help them predict the magnitude of products. In Activity Two students use what they know about multiplying by powers of 10 to help them predict the magnitude of products and place the decimal point in products.

Activity Three gives students an opportunity to use estimation to predict the magnitude of products and place the decimal point in products. Use one or more of these activities with your students, as needed.

TEACHING THE *INTERVENTION WORKSHOP* LESSON

Getting Started

Write the following on the board.

1.78 × 37

◆ **Say:** *Round each factor to its greatest place. What is 1.78 rounded to its greatest place?* (2) *What is 37 rounded to the greatest place?* (40) *How much is 2 times 40?* (80) *About how much is 1.78 times 37?* (80)

ACTIVITY ONE: Use Number Sense to Predict the Magnitude of Decimal Products

Write this exercise on the board.

0.483 × 80

Tell students that they can use what they know about the properties of multiplication and the meaning of decimals to predict about how great the product will be.

◆ **Say:** *Is 80 being multiplied by a number that is greater than 1 or less than 1?* (less than 1) *Will the product be greater than 80 or less than 80?* (less than 80) *How do you know?* (Possible response: 1 × 80 = 80; 0.483 is less than 1. So 0.483 × 80 will have a product that is less than 80.) *Is 0.483 closer to 1/2 or closer to 1?* (1/2) *Will the product be closer to 80 or to 40?* (40) *How do you know?* (0.483 is closer to 1/2 than to 1; 1/2 of 80 is 40.)

Write this exercise on the board.

4.83 × 80

◆ **Say:** *Is 80 being multiplied by a number that is greater than 1 or less than 1?* (greater than 1) *Will the product be greater than 80 or less than 80?* (greater than 80) *How do you know?* (Possible response: 1 × 80 = 80. 4.83 is greater than 1, so 4.83 × 80 will have a product that is greater than 80.) *Will the product be closer to 400 or to 4000?* (400) *How do you know?* (Possible response: 4.83 is a little less than 5. 5 × 80 = 400, so 4.83 × 80 is much less than 4,000.)

Have students work independently to complete the following exercises.

- Which is closest to the product of 0.81 × 12: 6, 10, 60, or 120? (10)
- Which is closest to the product of 0.53 × 200: 10, 20, 100, or 200? (100)
- Which is closest to the product of 0.87 × 62: 6.2, 62, 620, or 6200? (62)
- Which is closest to the product of 5.6 × 30: 1.8, 18, 180, or 1800? (180)
- Which is closest to the product of 42.85 × 30: 12, 120, 1200, or 12000? (1200)

ACTIVITY TWO: Use Powers of 10 to Predict the Magnitude of Decimal Products

Explain that students can use what they know about multiplying decimals by 10, 100, and 1000 to help them predict about how great the product of decimals will be.

Write the following on the board.

$$10 \times 0.641 = 6.41$$

$$100 \times 0.641 = 64.1$$

$$1000 \times 0.641 = 641$$

◆ **Say:** *How does the decimal point move when you multiply a decimal by 10?* (It moves 1 place to the right.) *How does the decimal point move when you multiply a decimal by 100?* (It moves 2 places to the right.) *How does the decimal point move when you multiply a decimal by 1000?* (It moves 3 places to the right.)

Write the following on the board.

$$74 \times 0.641 = 47434$$

Explain that the decimal point has not yet been placed in the product and that students will do so now.

◆ **Say:** *74 is between 10 and 100. What is 10 times 0.641?* (6.41) *What is 100 times 0.641?* (64.1) *Where would you place the decimal point in the product to make the product between 6.41 and 64.1?* (after the 7) *What is the product?* (47.434)

Write the following on the board.

$$874 \times 3.6 = 31464$$

Remind students that the decimal point has not yet been placed in the product.

◆ **Say:** *874 is between 100 and 1000. What is 100 times 3.6?* (360) *What is 1000 times 3.6?* (3600) *Where would you place the decimal point in the product to make the product between 360 and 3600?* (after the 6) *What is the product?* (3146.4)

Write the following on the board.

$$8 \times 0.794 = 6352$$

Remind students that the decimal point has not yet been placed in the product.

◆ **Say:** *8 is between 1 and 10. What is 1 times 0.794?* (0.794) *What is 10 times 0.794?* (7.94) *Where would you place the decimal point in the product to make the product between 0.794 and 7.94?* (after the 6) *What is the product?* (6.352)

Write the following on the board.

$$23 \times 4.89 = 11247$$

Remind students that the decimal point has not yet been placed in the product.

◆ **Say:** *23 is between 10 and 100. What is 10 times 4.89?* (48.9) *What is 100 times 4.89?* (489) *Where would you place the decimal point in the product to make the product between 48.9 and 489?* (after the 2) *What is the product?* (112.47)

Have students use what they know about multiplying by 10, 100, or 1000 to place the decimal point in the product in each of the following examples.

$$94 \times 0.59 = 5546 \ (55.46)$$

$$17 \times 3.781 = 64277 \ (64.277)$$

$$7 \times 48.36 = 33852 \ (338.52)$$

$$206 \times 46.35 = 95481 \ (9548.1)$$

ACTIVITY THREE: Use Estimation to Predict the Magnitude of Decimal Products

Write the following on the board.

$$17 \times 5.63 = 9571$$

Tell students that they can use what they know about rounding and estimating to place the decimal point in the product.

◆ **Say:** *To estimate a product you can round each factor to its greatest place, then multiply. What is 17 rounded to its greatest place?* (20) *What is 5.63 rounded to its greatest place?* (6) *What is 20 times 6?* (120) *In the original example, 17 times 5.63,*

where would you place the decimal point so that the product was close to 120? (after the 5) *What is the product?* (95.71)

Have students use what they know about estimating products to place the decimal point in each of the following examples.

$$4 \times 7.385 = 2954 \ (29.54)$$

$$32 \times 2.6 = 832 \ (83.2)$$

$$79 \times 5.004 = 395316 \ (395.316)$$

SUMMING-IT-UP QUIZ

Give this quiz to students.

Write the following on the board.

$$6.73 \times 80$$

$$0.613 \times 18$$

◆ **Say:** *Which is closest to the product of 6.73 × 80: 5.6, 56, 560, or 5600?* (560) *Which is closest to the product of 0.613 × 18: 9, 18, 90, or 180?* (9)

Write the following on the board.

$$21 \times 8.731 = 183351$$

◆ **Say:** *How can you use what you know about multiplying with 10, 100, and 1000 to place the decimal point in the product?* (Since 21 is between 10 and 100, find 10 × 8.731 = 87.31, find 100 × 8.731 = 873.1, and place the decimal point so that the product is between 87.31 and 873.1)

◆ **Say:** *How can you use estimation to place the decimal point in the product?* (Estimate 20 × 9 = 180 and place the decimal point so that the product is as close as possible to that estimate.)

If students answer the *Summing-It-Up Quiz* correctly, they should then resume work on page 127 of their textbook. If students do not successfully complete the *Summing-It-Up Quiz*, further remediation may be necessary.

Multiplication: Placing the Decimal Point

Use with Lesson 4-5, exercises 1–15, text pages 130–131.

DIAGNOSTIC INTERVIEW

Before beginning this *Intervention Workshop* lesson, have students complete the following exercises.

Diagnostic Exercises

◆ **Say:** *Write these decimal multiplication exercises. Then multiply to find each product.*

$$5 \times 0.7 \qquad 8 \times 6.2 \qquad 0.3 \times 9.4$$

(3.5; 49.6; 2.82)

◆ **Say:** *Write 57.0067 and tell how many decimal places it has.* (4 decimal places)

◆ **Say:** *Write 670892 and place the decimal point to show 5 decimal places.* (6.70892)

If students answer the Diagnostic Exercises correctly, there is probably no need for them to do this *Intervention Workshop* lesson. Have those students rejoin their class on page 131 of the student textbook. For students who have had difficulty answering the exercises above, continue the *Intervention Workshop*.

BACKGROUND

This *Intervention Workshop* lesson addresses the problem of where to place the decimal point in a decimal product and why counting the number of decimal places in the factors provides the correct number of decimal places in the product. Students begin by multiplying a decimal by a whole number using base ten grid models. Counting the shaded columns and/or squares on the base ten grids provides a clear visual methodology for students to readily determine the decimal product. Next students multiply two decimals using this same methodology. At the end of the lesson students count the number of decimal places in a given decimal.

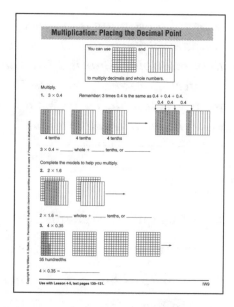

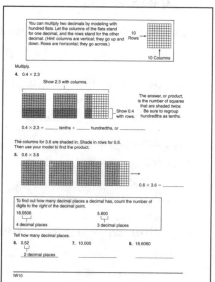

TEACHING THE *INTERVENTION WORKSHOP* LESSON

Getting Started

Materials: centimeter grid paper (BLM 1)

◆ **Say:** *Remember that another way to say 6 times 13 is to say six 13s. This means that you can show multiplication by repeated addition. For this example, 6 times 13 equals 78 is the same as 13 plus 13 plus 13 plus 13 plus 13 plus 13 equals 78. Notice that when you multiply 6 times 13 your answer is the same as when you add six 13s.*

◆ **Say:** *Write these multiplication exercises.*

    4 × 15   3 × 42   2 × 95

Write each multiplication as a repeated addition, then find the product and sum. Are each sum and product equal? Remember: The product *is the answer to a multiplication and the* sum *is the answer to an addition.* (15 + 15 + 15 + 15, 60; 42 + 42 + 42, 126; 95 + 95, 190; Yes, the sum and product in each case are equal.)

ⓑ USING PAGES IW9 and IW10

Have a student read aloud the material in the box at the top of page IW9. Then have a student read aloud the direction line above exercise 1 and the *Remember* note to the right of the exercise. Relate the *Remember* note to the exercises students completed earlier in *Getting Started* in the following way.

◆ **Say:** *Predict whether you think the product and sum in exercise 1 will be equal. Tell why.* (The product and sum will be equal because multiplication is the same as repeated addition.)

Exercises 1–3 on page IW9
Assist students to complete exercise 1 the following way.

◆ **Say:** *The model to the right of the arrow is a combination of the model to the left of the arrow. The model on the left shows the multiplication as repeated addition, so the two models show the same thing. You count the number of shaded columns in the model on the right to find the answer to the multiplication.*

Then have a student read aloud the direction line above exercises 2–3.

◆ **Say:** *In these two exercises you complete the base ten grid models by shading in the missing columns or squares to find the products.*

By using base ten grid models and counting the shaded columns and/or squares, these three exercises lead students to determine the product of a whole number and a decimal.

Exercises 4–8 on page IW10
Before beginning exercise 4, you may want to have students complete this interim exercise: 3 × 1.72

◆ **Say:** *Draw as many 10 by 10 grids on grid paper as you need in order to find the answer to 3 times 1.72. Shade in as many squares as you need to show the decimals just as you did in exercise 3.*

Have a student read aloud the material in the box at the top of page IW10.

◆ **Say:** *When you multiply two decimals, you can use area to show the product. On a 10 by 10 grid you show one decimal of the multiplication with shaded columns and the other decimal with rows shaded on the columns. In this way you easily see that the area that is shaded twice shows the answer.*

◆ **Say:** *Remember that rows go across and columns go up and down. When you shade in your model, it does not matter which decimal you show with columns and which decimal you show with rows.*

Assist students as necessary to complete exercise 4. Then have a student read aloud the direction line above exercise 5.

◆ **Say:** *In this exercise you will shade the rows to show 0.6. Remember that the area that is shaded twice shows the answer.*

For students who have difficulty with exercises 4 and 5, use the following additional examples [additional example for exercise 4: 0.3 × 0.8; for exercise 5: 1.2 × 2.4 (0.24, 2.88)].

Have a student read aloud the material in the box below exercise 5.

◆ **Say:** *You need to know how to find the number of decimal places in a decimal. This is important so that you can apply the rule of counting and adding the number of decimal places in the factors when you are ready to place the decimal point correctly in the product.*

◆ **Say:** *Remember: A factor is a number you multiply.*

Have a student read aloud the direction line above exercises 6–8. Assist students as necessary to complete these exercises.

◆ **Say:** *Look back at exercises 1–3 on page 9. Count the number of decimal places in the decimal in exercise 1. Then count the number of decimal places in the product, or answer to exercise 1. What can you say about the number of decimal places you counted?* (There is 1 decimal place in the factor and 1 decimal place in the product. They are the same.)

Guide students to analyze the factors and products in exercises 2 and 3 in the same manner.

◆ **Say:** *Now look back at exercises 4 and 5 on page 10. Count the number of decimal places in the decimals in exercise 4. Then count the number of decimal places in the product or answer to exercise 4. What can you say about the number of decimal places you counted?* (There is 1 decimal place in each factor and 2 decimal places in the product. The sum of the number of decimal places in the factors is the same as the number of decimal places in the product.)

You can guide students to analyze exercise 5 the same way. Assist them as necessary to conclude that the sum of the decimal places in the factors is equal to the number of decimal places in the product.

Read aloud the material in the box following exercise 5. Make sure students understand the sample answer for exercise 6. Then have students complete exercises 7 and 8.

SUMMING-IT-UP QUIZ

Give this quiz to students.

◆ **Say:** *Write these decimal multiplication exercises. Find each product.*

$$4 \times 0.2 \qquad 3 \times 1.34 \qquad 0.2 \times 1.9$$

(0.8; 4.02; 0.38)

◆ **Say:** *Write the decimal 10.0023 and tell how many decimal places it has.* (4 decimal places)

◆ **Say:** *Write a decimal that has 5 decimal places.* (sample answer: 0.02109)

If students answer the *Summing-It-Up Quiz* correctly, they should return to Lesson 4-5 and resume work on page 131 of their textbook. If students do not successfully complete the *Summing-It-Up Quiz*, further remediation may be necessary.

Answers: Pages IW9 and IW10

1. 1; 2; 1.2

2. 3; 2; 3.2

3. 1.40

4. 9; 2; 0.92

5. 2.16

6. 2 decimal places

7. 3 decimal places

8. 4 decimal places

Decimals Divided by Whole Numbers

Use with Lesson 4-9, exercises 3–18, text pages 138–139.

DIAGNOSTIC INTERVIEW

Before beginning this *Intervention Workshop* lesson, have students complete the following exercises.

Diagnostic Exercises

Write this exercise on the board.

$$30\overline{)12.6}$$

◆ **Say:** *Will this quotient be greater than 1 or less than 1?* (less than 1) *How do you know?* (Possible response: Since 12.6 is less than 30, there is less than one group of 30 in 12.6.) *When you divide a decimal by a whole number, when will the quotient be less than 1?* (when the dividend is less than the divisor) *When will the quotient be greater than 1?* (when the dividend is greater than the divisor)

Write this exercise on the board.

$$6\overline{)0.186}$$

◆ **Say:** *Is the correct quotient 0.31 or 0.031?* (0.031) *How do you know?* (Possible response: When you divide a decimal by a whole number, the quotient is less than the dividend. Since 0.31 is greater than 0.186, it cannot be the quotient. Therefore 0.031 must be the quotient.)

If students answer the Diagnostic Exercises correctly, there is probably no need for them to do this *Intervention Workshop* lesson. Have those students rejoin their class on page 139 of the student textbook. For students who have had difficulties answering the exercises above, continue the *Intervention Workshop*.

BACKGROUND

Exercises 3–18 on page 139 of the student textbook require students to divide decimals by 1- and 2-digit whole numbers. These *Intervention Workshop* activities address the needs of students who still have not mastered using number sense and estimation to predict the magnitude of decimal quotients, placing a decimal point in a quotient, and checking the reasonableness of answers. In Activity One students use what they know about dividing with money to help them predict the magnitude of decimal quotients. In Activity Two students use number sense and a place-value frame to help them predict the magnitude of decimal quotients. Activity Three gives students an opportunity to use estimation to predict the magnitude of quotients and place the decimal point in quotients. Use one or more of these activities with your students, as needed.

TEACHING THE *INTERVENTION WORKSHOP* LESSON

Getting Started

Write the following on the board.

$$4\overline{)3.62}$$

◆ **Say:** *What is the dividend?* (3.62) *What is the divisor?* (4) *Is the dividend greater than or less than the divisor?* (The dividend is less than the divisor.) *Can you divide when the dividend is less than the divisor?* (Yes)

ACTIVITY ONE: Use Money to Predict the Magnitude of Decimal Quotients

Materials: place-value frames for tens, ones, tenths, and hundredths (BLM 6) and for $10, dollars, dimes, and pennies (BLM 7)

Write the following on the board.

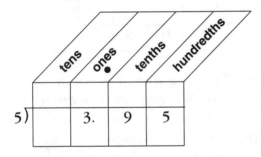

◆ **Say:** *How are the two problems alike?* (The numbers in each place are the same.) *How are the problems different?* (In one problem ones, tenths, and hundredths are divided by a whole number; in the other problem, dollars, dimes, and pennies are divided by a whole number.)

Have students write the problems in place-value frames. Then guide students to predict the magnitude of the money quotient.

◆ **Say:** *When you divide $3.95 by 5, will the quotient be greater or less than $1?* (less than $1) *How do you know?* (possible response: 5 × $1 = $5, and $3.95 is less than $5.) *Will you have less than 10 cents?* (No) *How do you know?* (5 × 10¢ is only 50¢, and $3.95 is greater than 50¢.) *The quotient is less than $1.00, but greater than 10 cents. In what place will the quotient begin?* (the dimes place) *Which could be the quotient: $7.90 or $0.79?* ($0.79)

Have students use the money quotient to help them find the corresponding decimal quotient.

◆ **Say:** *In what place does the quotient begin: tens, ones, tenths, or hundredths?* (tenths) *What is the quotient?* (0.79)

Write the following on the board.

$$7\overline{)41.86} \qquad 7\overline{)\$41.86}$$

Have students write the problems in place-value frames. Then guide students to predict the magnitude of the money quotient.

◆ **Say:** *When you divide $41.86 by 7, will the quotient be greater or less than $10?* (less than $10) *How do you know?* (Possible response: 7 × $10 = $70, and

$41.86 is less than $70.) *Will the quotient be greater or less than $1?* (greater than $1) *How do you know?* (Possible response: 7 × $1 = $7, and $41.86 is greater than $7.) *The quotient is less than $10, but greater than $1. In what place will the quotient begin?* (the dollars place) *Which could be the quotient: $59.80, $5.98, or $0.598?* ($5.98)

Have students use the money quotient to help them find the corresponding decimal quotient.

◆ **Say:** *In what place does the quotient begin: tens, ones, tenths, or hundredths?* (ones) *What is the quotient?* (5.98)

ACTIVITY TWO: Use a Place-Value Frame and Number Sense to Predict the Magnitude of Decimal Quotients

Materials: place-value frames for tens, ones, tenths, hundredths, thousandths, and ten-thousandths (BLM 8)

Write the following on the board.

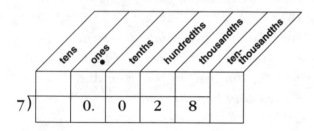

◆ **Say:** *Will the quotient begin in the ones place?* (No) *How do you know?* (There are no ones to divide.) *Will the quotient begin in the tenths place?* (No) *How do you know?* (There are no tenths to divide.) *How many hundredths are there?* (2) *Will the quotient begin in the hundredths place?* (No) *How do you know?* (Possible responses: Since 2 is less than 7, there are not enough hundredths to divide; multiplying 7 × 0.01 would give you 0.07, which is greater than 0.02.) *Will the quotient begin in the thousandths place?* (Yes) *How do you know?* (Possible response: Since 28 is greater than 7, there are enough thousandths to divide.) *Which could be the quotient: 4, 0.4, 0.04, 0.004, or 0.0004?* (0.004)

Write the following on the board.

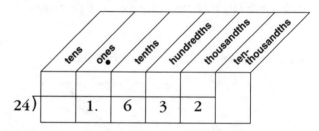

◆ **Say:** *Will the quotient begin in the ones place?* (No) *How do you know?* (Possible responses: Since 1 is less than 24, there are not enough ones to divide; multiplying 24 × 1 would give you 24, which is greater than 1.) *Will the quotient begin in the tenths place?* (No) *How do you know?* (Possible responses: Since 16 is less than 24, there are not enough tenths to divide; multiplying 24 × 0.1 would give you 2.4, which is greater than 1.6) *Will the quotient begin in the hundredths place?* (Yes) *How do you know?* (Possible response: Since 163 is greater than 24, there are enough hundredths to divide.) *Which could be the quotient: 6.8, 0.68, 0.068, or 0.0068?* (0.068)

Have students use a place-value frame to complete the following examples.

• Which could be the quotient of 13.02 ÷ 42: 3.1, 0.31, or 0.031? (0.31)

- Which could be the quotient of 58.667 ÷ 7: 8.381, 0.831, or 0.0831? (8.381)

- Which could be the quotient of 0.0828 ÷ 9: 0.92, 0.092, or 0.0092? (0.0092)

ACTIVITY THREE: Use Estimation to Predict the Magnitude of Decimal Quotients

Write the following on the board.

$$41.352 \div 8 = 5169$$

Tell students that they can use what they know about compatible numbers and estimating to place the decimal point in the quotient. Remind students that to find compatible numbers they should think of numbers that divide evenly, such as $10 \div 2 = 5$ or $210 \div 7 = 30$.

◆ Say: *What compatible numbers can you use for 41.352 divided by 8?* ($40 \div 8$) *What is 40 divided by 8?* (5) *In the sequence of digits 5169, where would you place the decimal point so that the quotient is close to 5?* (after the 5) *What is the quotient?* (5.169)

Have students use what they know about estimating quotients to place the decimal point in each of the following examples.

$$55.962 \div 9 = 6218 \ (6.218)$$
$$246.444 \div 6 = 41074 \ (41.074)$$
$$8.701 \div 7 = 1243 \ (1.243)$$

SUMMING-IT-UP QUIZ

Give this quiz to students.

Write the following on the board.

$$3.198 \div 26$$
$$0.602 \div 7$$
$$92.48 \div 34$$

◆ Say: *Which could be the quotient of 3.198 divided by 26: 1.23, 0.123, or 0.0123?* (0.123) *Which could be the quotient of 0.602 divided by 7: 8.6, 0.86, or 0.086?* (0.086) *Which could be the quotient of 92.48 divided by 34: 2.72, 0.272, or 0.0272?* (2.72)

Write the following on the board.

$$28.023 \div 3 = 9341$$

◆ Say: *How can you use estimation and compatible numbers to place the decimal point in the quotient?* (Estimate $27 \div 3 = 9$ and place the decimal point in the quotient so that the quotient is as close as possible to that estimate.)

If students answer the *Summing-It-Up Quiz* correctly, they should then resume work on page 139 of their textbook. If students do not successfully complete the *Summing-It-Up Quiz,* further remediation may be necessary.

Division: Placing the Decimal Point

Use with Lesson 4-10, exercises 5–32, text pages 140–141.

DIAGNOSTIC INTERVIEW

Before beginning this *Intervention Workshop* lesson, have students complete the following exercises.

Diagnostic Exercises

◆ **Say:** *Write these decimal division exercises. Then divide to find the quotient.*

$$4.5 \div 5 \qquad 7.2 \div 6 \qquad 2.16 \div 0.72$$

(0.9; 1.2; 3)

◆ **Say:** *Write these divisions. Then use a pattern to find the quotients.*

$$12{,}400 \div 400 = (31)$$
$$1240 \div 40 = (31)$$
$$124 \div 4 = (31)$$
$$12.4 \div 0.4 = (31)$$
$$1.24 \div 0.04 = (31)$$

◆ **Say:** *What does the pattern show?* (The quotients are the same.)

If students answer the Diagnostic Exercises correctly, there is probably no need for them to do this *Intervention Workshop* lesson. Have those students rejoin their class on page 141 of the student textbook. For students who have had difficulty answering the exercises above, continue the *Intervention Workshop*.

BACKGROUND

This *Intervention Workshop* lesson addresses the problem of where to place the decimal point in a quotient when dividing a decimal by a decimal and why making the divisor a whole number and moving the decimal point the same number of places in the dividend results in the correct number of decimal places in the quotient. Students begin with dividing a decimal by a whole number, using base 10 grid models. Counting the number of shaded columns and/or squares in equal groups on the base 10 grids provides a clear visual

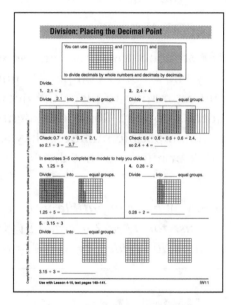

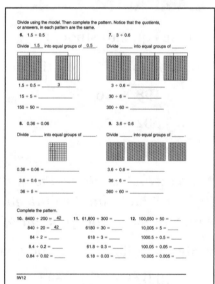

methodology for students to readily determine the quotient. Next students divide a decimal by a decimal using this same methodology. Also as part of these exercises they complete division patterns based on the quotient they find. At the end of the lesson students complete additional patterns to help them find the number of decimal places in the quotient after dividing a decimal by a decimal.

TEACHING THE *INTERVENTION WORKSHOP* LESSON

Getting Started

Materials: counters, centimeter grid paper (BLM 1)

◆ **Say:** *When you divide a whole number by a whole number you can show the division with counters. In the example 27 divided by 3 you take 27 counters and form 3 equal groups. The number of counters in each equal group is the* quotient, *or the answer, to 27 divided by 3. The* dividend, *27, is the number of counters you use and the* divisor, *3, is the number of equal groups you form.*

You may want to show students this array with counters or make a drawing to help students visualize what you are describing.

◆ **Say:** *Write these division exercises. Show each division with counters or make a drawing of the counters and find each quotient.*

    39 ÷ 3   *48 ÷ 6*   *28 ÷ 7*   (13; 8; 4)

◖ USING PAGES IW11 and IW12

Exercises 1–5 on page IW11
Have a student first read aloud the material in the box at the top of page IW11 then read aloud the direction line above exercises 1–2. As you assist students with exercise 1, point out the brackets above the model.

◆ **Say:** *Count the number of brackets. How many are there?* (3)

◆ **Say:** *The brackets will guide you to see the number of equal groups. Now count the number of shaded columns in each bracketed group. How many are there? What does this number of shaded columns represent?* (7; 7 shaded columns represent 0.7.)

◆ **Say:** *The decimal for the shaded part in each equal group is the quotient of the division. In exercise 1 what is the quotient?* (0.7)

Relate the methodology shown in this exercise to the whole-number divisions students modeled with counters earlier in *Getting Started* in the following way.

◆ **Say:** *How is finding a quotient of a decimal and a whole number using base 10 grids like finding the quotient of two whole numbers using counters?* (The quotient is the whole number or decimal in the equal group.)

Assist students as necessary to complete exercise 2. Have a student read aloud the direction line above exercises 3–5.

◆ **Say:** *In these three exercises you complete the base 10 grid models by shading in the missing columns or squares to show the dividend. Then you divide the shaded part into equal groups. Outline each equal group with a heavy marker so you can easily see the decimal that each equal group represents.*

By using base 10 grid models and counting the shaded columns and/or squares within each equal group, exercises 1–5 lead students to determine the quotient of a decimal divided by a whole number.

Exercises 6–9 on page IW12
Have a student read aloud the direction line above exercises 6–9.

In exercises 6–8 students use the models provided to find the quotient, while in exercise 9 they must outline the equal groups on the models to find the quotients. Within the exercises students complete patterns based on the quotient they find.

◆ **Say:** *In exercise 9 you must make your own model by shading the grids and outlining the equal groups.*

Assist students as necessary to complete exercise 9.

If in exercises 8 and 9 students have difficulty answering the division-by-a-decimal problems in the pattern, have them skip these and answer the whole-number divisions. Have students return to the decimal divisions they skipped and find these quotients based on the pattern they discovered.

◆ **Say:** *Look back at the patterns you completed in exercises 6–9. What can you say about these patterns?* (The quotients, or answers, are always the same in each pattern.)

Exercises 10–12 on page IW12
Have a student read aloud the direction line above exercises 10–12. Guide students as necessary to complete exercises 10–12.

◆ **Say:** *What can you say about these patterns?* (You can use the answers to whole-number division patterns to find the answers to decimal divisions in the same pattern.)

SUMMING-IT-UP QUIZ

Give this quiz to students.

◆ **Say:** *Write these decimal division exercises. Find each quotient.*

$7.2 \div 9$ $0.32 \div 8$ $3.3 \div 1.2$

(0.8; 0.04; 2.75)

◆ **Say:** *Write this pattern and then complete it.*

$33{,}800 \div 2600 = (13)$

$3380 \div 260 = (13)$

$338 \div 26 = (13)$

$33.8 \div 2.6 = (13)$

$3.38 \div 0.26 = (13)$

If students answer the *Summing-It-Up Quiz* correctly, they should return to Lesson 4-10 and resume work on page 141 of their textbook. If students do not successfully complete the *Summing-It-Up Quiz*, further remediation may be necessary.

Answers: Pages IW11 and IW12

2. 2.4; 4; 0.6

3. 1.25; 5; 0.25

4. 0.28; 2; 0.14

5. 3.15; 3; 1.05

6. 3; 3

7. 3; 0.6; 5; 5; 5

8. 0.36; 0.06; 6; 6; 6

9. 3.6; 0.6; 6; 6; 6

10. 42; 42; 42

11. 206; 206; 206; 206; 206

12. 2001; 2001; 2001; 2001; 2001

Understanding Factoring

Use with Lesson 5-4, exercises 12–21, text pages 168–169.

DIAGNOSTIC INTERVIEW

Before beginning this *Intervention Workshop* lesson, have students complete the following exercises.

Diagnostic Exercises

◆ **Say:** *Write these numbers: 16, 29, 35. Tell which is a prime number and which is a composite number. Then tell why.* (16: composite; it has factors other than 1 and 16; 29: prime; it has only 1 and 29 as factors; 35: composite; it has factors other than 1 and 35)

◆ **Say:** *Write 46 and 72 as a product of their prime factors.* (46 = 2 × 23; 72 = 2 × 2 × 2 × 3 × 3)

If students answer the Diagnostic Exercises correctly, there is probably no need for them to do this *Intervention Workshop* lesson. But before they resume work on page 169, you may want to assign the review lesson on page 4 of *Skills Update* in the student textbook. Then have students rejoin their class on page 169. For students who have had difficulty, continue this *Intervention Workshop*.

BACKGROUND

This *Intervention Workshop* lesson addresses the problem of the distinction between a prime number and a composite number. Students begin with drawing as many rectangular arrays as possible for a number. Discovering whether there is only one array or more than one array for a number provides a clear visual methodology for students to distinguish a prime number from a composite number. In preparation for writing a number as a product of its prime factors, students then find all of the prime numbers to 100, using the Sieve of Eratosthenes. Finally students use factor trees to find the prime factorization of numbers.

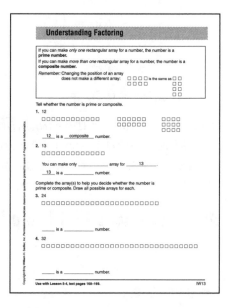

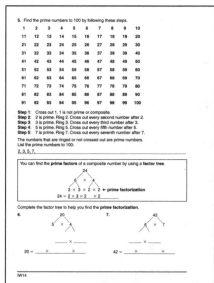

TEACHING THE *INTERVENTION WORKSHOP* LESSON

Getting Started

Materials: counters, centimeter grid paper (BLM 1)

◆ **Say:** *One number is divisible by another if you can divide and there is no remainder. Write any number divisible by 2.* (sample answer: 22) *Write any number divisible by 5.* (sample answer: 35) *Write any number divisible by 10.* (sample answer: 100)

◆ **Say:** *When you count by 2s you name even numbers. Write a rule about the divisibility of even numbers by 2.* (All even numbers are divisible by 2.)

◆ **Say:** *When you count by 5s, you name numbers divisible by what number?* (by 5) *When you count by 10s, you name numbers divisible by what number(s)?* (by 2, 5, and 10)

Assist students as necessary to find all of the numbers that evenly divide given numbers ending in zero by reminding them that numbers ending in zero are divisible by 10 and that numbers divisible by 10 are also divisible by 2 and 5.

USING PAGES IW13 and IW14

Have a student read aloud the material in the box at the top of page IW13.

◆ **Say:** *A rectangular array is an arrangement of tiles, counters, buttons, and so on, placed in rows and columns. There are an equal number of tiles in each row and an equal number of tiles in each column. If you interchange the number of rows with the number of columns, you do not create a different array. You can see a drawing of this in the box. These two arrays are the same.*

◆ **Say:** *Remember: Rows go across and columns go up and down.*

Exercises 1–4 on page IW13
Have a student read aloud the direction line above exercises 1-2. In these exercises pictures of tile models are provided for students to interpret.

Next have a student read aloud the direction line above exercises 3-4.

In these exercises students draw their own arrays for the given numbers. Guide students as necessary to draw all possible arrays for each number.

By using or drawing arrays these four exercises lead students to analyze whether the given number is prime or composite.

◆ **Say:** *You may want to use counters to show all possible arrays for the numbers in exercises 3 and 4. This will help you decide whether you have created a different array or the same array. Remember that when you interchange the number of rows with the number of columns you have not created a different array.*

Exercises 5–7 on page IW14
Have a student read aloud the direction line in exercise 5. Assist students as necessary to follow the steps given below the hundred chart.

◆ **Say:** *Look at exercise 1 on page IW13. When you count the number of rows and the number of columns in an array for a number, the two numbers you count are factors of the number. Remember that a number you multiply is a* factor.

◆ **Say:** *Look at exercise 1. Count and then write the number of rows and columns in the first array.* (1 row and 12 columns) *These two numbers are factors of 12. What are the factors of 12 shown in the first array?* (1 and 12)

You can guide students to find the remaining factors of 12 from arrays in exercise 1 the same way.

◆ **Say:** *If you write all of the numbers in order that you counted from the rows and columns of the arrays, you will create a list of all of the factors of 12. Now write all the factors of 12.* (1, 2, 3, 4, 6, 12)

If necessary you can guide students to name all the factors of 13, 24, and 32 in exercises 2–4 in the same way.

Return students' attention to page IW14 and have a student read the material in the box in the middle of the page. Then have a student read aloud the direction line above exercises 6–7.

◆ **Say:** *Prime factorization means that you write a number as a product of its prime factors. You can use the list of prime numbers you found in exercise 5 to help you decide whether the numbers you write on the bottom branches of your factor tree are prime.*

Assist students as necessary to write the prime factorization in exercises 6–7 by pointing out that the prime factors shown on the bottom branches of the factor tree are written in order for the prime factorization.

SUMMING-IT-UP QUIZ

Give this quiz to students.

◆ **Say:** *Tell whether the number is prime or composite. Explain why.*

   23   77

(23: prime; it only has 1 and 23 as factors; 77: composite; it has 7 and 11, as well as 1 and 77, as factors)

◆ **Say:** *Write the prime factorization of 80 and 216.* (80 = 2 × 2 × 2 × 2 × 5; 216 = 2 × 2 × 2 × 3 × 3 × 3)

If students answer the *Summing-It-Up Quiz* correctly, you may want to assign the review lesson on page 4 of *Skills Update* in the student textbook. Then students should return to Lesson 5-4 and resume work on page 169 of their textbook. If students do not successfully complete the *Summing-It-Up Quiz*, further remediation may be necessary.

Answers: Pages IW3 and IW4

2. 1; prime

3. 1 by 24; 2 by 12; 3 by 8; 4 by 6; 24; composite

4. 1 by 32; 2 by 16; 4 by 8; 32; composite

5. primes greater than 7: 11, 13, 17, 19, 23, 29, 31, 37, 41, 43, 47, 53, 59, 61, 67, 71, 73, 79, 83, 89, 97

6. 2; 2; 5 × 2 × 2

7. 2; 3; 2 × 3 × 7

Greatest Common Factor

Use with Lesson 5-6, exercises 25–48, text pages 172–173.

DIAGNOSTIC INTERVIEW

Before beginning this *Intervention Workshop* lesson, have students complete the following exercises.

Diagnostic Exercises

◆ **Say:** *Write 32 and 48 and find all of the factors of each.* (factors of 32: 1, 2, 4, 8, 16, 32; factors of 48: 1, 2, 3, 4, 6, 8, 12, 16, 24, 48)

◆ **Say:** *Write these pairs of numbers. Find the common factors and then the greatest common factor of each pair.*

10 and 35   36 and 81

(common factors: 1, 5; GCF: 5; common factors: 1, 3, 9; GCF: 9)

If students answer the Diagnostic Exercises correctly, there is probably no need for them to do this *Intervention Workshop* lesson. Have those students rejoin their class on page 173 of the student textbook. For students who have had difficulty answering the exercises above, continue the *Intervention Workshop*.

BACKGROUND

This *Intervention Workshop* lesson addresses the relationship between the common factors and the greatest common factor of two numbers; understanding this relationship is a precursor to the simplification of fractions. Students begin by drawing all possible rectangles on grid paper to show all the factors of a number. Drawing rectangles provides a clear visual model for students to ensure that they list all the factors of a number. Then they draw all possible rectangles for two numbers to help them find the common factor(s) and the greatest common factor of the two numbers.

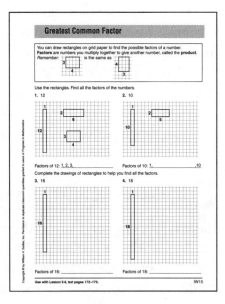

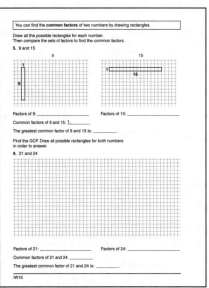

TEACHING THE *INTERVENTION WORKSHOP* LESSON

Getting Started

Materials: centimeter grid paper (BLM 1)

◆ **Say:** *Write these pairs of multiplication expressions.*

4×5 and 5×4 10×3 and 3×10 1×7 and 7×1

◆ **Say:** *Are results the same for each pair? Tell why.* (Yes; changing the order in which numbers are multiplied does not change the answer.)

◆ **Say:** *Factors are numbers multiplied to find a product. You can divide to find if one number is a factor of another. Write the following questions and explain your answer to each.*

Is 7 a factor of 35?   Is 12 a factor of 23?

(Yes; because $7 \times 5 = 35$ and 7 divides evenly into 35, $7\overline{)35}^{\,5}$; No; because there is no whole-number replacement for $12 \times \underline{\ ?\ } = 23$ and 12 does not divide 23 evenly.)

⊘ USING PAGES IW15 and IW16

Have a student read aloud the material in the box at the top of page IW15. Relate the two rectangles shown to the exercises students completed earlier in *Getting Started* in the following way.

◆ **Say:** *Look at the first rectangle in the box. Count the squares inside. Now look at the second rectangle and count the squares inside it. What did you count and what can you say about the two numbers?* (12; The numbers are the same.)

◆ **Say:** *When you look at the two numbers written along the sides of each rectangle, you see 3 and 4. When you counted the squares inside each rectangle as 12, you can see that 3 times 4 is the same as 4 times 3 because both equal 12. This tells you that the two rectangles have the same area.*

◆ **Say:** *The numbers 3 and 4 are the length and width of the rectangle, respectively. They are factors of 12 because 3 times 4 equals 12. Remember: 12 is the number of squares inside the rectangle.*

Exercises 1–4 on page IW15

Have a student read aloud the direction line above exercises 1-2. In these exercises the rectangles for each number are provided for students to interpret. These exercises guide students to find and then write all the factors of a number in order, from least to greatest.

◆ **Say:** *You write the factors in order, from least to greatest. This will make it easier to compare the factors of two numbers and to find their common factors.*

If students have difficulty with exercises 1-2, have them count the number of squares inside each rectangle. This will show them that they are looking for all of the factors of the same number.

Next have a student read aloud the direction line above exercises 3-4. In these exercises students draw their own rectangles for the given numbers. Guide students as necessary to draw all possible rectangles for each number.

◆ **Say:** *In exercise 3 remember that you are going to draw all possible rectangles that have 16 squares inside.*

You can assist students in the same manner with exercise 4. By using or drawing rectangles, these four exercises lead students to list all of the factors of a number in order.

It is possible that some students may confuse listing all the factors of a number with writing the prime factorization of a number. If this occurs stress that finding all factors of a number is a *list* of all factors, prime and composite. The prime factorization of a number is a *product* of prime factors.

Exercises 5–6 on page IW16
Have a student read aloud the sentence in the box at the top of page IW16. Then have the student read aloud the direction line above exercise 5.

◆ **Say:** *You will find the common factors of the two numbers by drawing all possible rectangles. Because you will write the factors in order, from least to greatest, it will be easier for you to see the greatest common factor of the two numbers. Remember: The greatest common factor is the greatest factor common to both numbers. It will be the greatest number common to each of your two lists of numbers.*

Have a student read aloud the direction line above exercise 6. Remind students that *GCF* stands for *greatest common factor*. Guide students as necessary to draw all possible rectangles, write the list of factors in order, and identify the common factors and the greatest common factor.

For students who have difficulty with exercises 5-6 use the following additional examples for them to draw rectangles and find the greatest common factor [additional examples: 8 and 40; 30 and 75 (8; 15)].

SUMMING-IT-UP QUIZ

Give this quiz to students.

◆ **Say:** *Write 49 and 82 and write all of the factors of each in order from least to greatest.* (factors of 49: 1, 7, 49; factors of 82: 1, 2, 41, 82)

◆ **Say:** *Write these pairs of numbers. What is the greatest common factor of each pair of numbers?*

    39 and 65   105 and 150   (GCF of 39 and 65: 13; GCF of 105 and 150: 15)

◆ **Say:** *Paul says the greatest common factor of 90 and 108 is 9. Mimi says the greatest common factor of 90 and 108 is 18. Who is correct?* (Mimi)

If students answer the *Summing-It-Up Quiz* correctly, then they should return to Lesson 5-6 and resume work on page 173 of their textbook. If students do not successfully complete the *Summing-It-Up Quiz,* further remediation may be necessary.

Answers: Pages IW15 and IW16

1. 4, 6, 12   **2.** 2, 5

4. 2 by 9; 3 by 6; 1, 2, 3, 6, 9, 18

6. 1 by 21; 3 by 7; factors of 21: 1, 3, 7, 21; 1 by 24; 2 by 12, 3 by 8; 4 by 6; factors of 24: 1, 2, 3, 4, 6, 8, 12, 24; common factors: 1, 3; GCF: 3

3. 2 by 8; 4 by 4; 1, 2, 4, 8, 16

5. factors of 9: 1, 3, 9; factors of 15: 1, 3, 5, 15; common factors: 3; GCF: 3

Finding Equivalent Fractions

Use with Lesson 6-3, exercises 1–17, text pages 206–207, and Lesson 6-5, exercises 13–16, text pages 210–211.

DIAGNOSTIC INTERVIEW

Before beginning this *Intervention Workshop* lesson, have students complete the following exercises.

Diagnostic Exercises

◆ **Say:** *Write 3/8, 5/9, and 35/40. Write any three equivalent fractions for each.* (sample answers: 6/16, 9/24, 12/32; 10/18, 15/27, 20/36; 7/8, 14/16, 70/80)

◆ **Say:** *Write these exercises. Add or subtract. Then write your answer in simplest form.*

     5/8 + 1/8   1/3 + 1/6   3/5 − 2/10
(3/4; 1/2; 2/5)

If students answer the Diagnostic Exercises correctly, there is probably no need for them to do this *Intervention Workshop* lesson. But before they resume work on page 207 or 211, you may want to assign the review lessons on pages 15 and 16 of *Skills Update* in the student textbook. Then have students rejoin their class on page 207 or 211. For students who have had difficulty, continue this *Intervention Workshop*.

BACKGROUND

This *Intervention Workshop* lesson addresses the problem of the relationship between finding equivalent fractions and adding and subtracting fractions. Finding equivalent fractions is a necessary skill for adding and subtracting fractions with unlike denominators and for writing fractions in simplest form. Students begin with analyzing equal rectangles on grid paper that show equivalent fractions for a given fraction. Comparing the sizes of the rectangles and the number of equal parts provides a clear visual model to help students understand equivalent fractions. Then students analyze fraction tapes on which are written equivalent fractions in order of multiples of their terms. Finally they use fraction strips to add and subtract fractions with like and unlike denominators and to write the answers in simplest form.

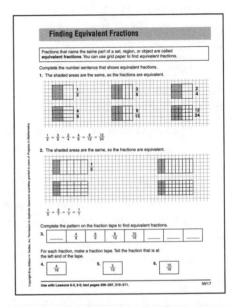

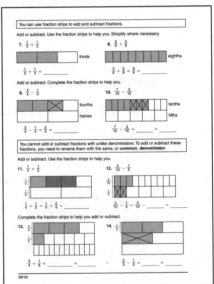

TEACHING THE *INTERVENTION WORKSHOP* LESSON

Getting Started

Materials: centimeter grid paper (BLM 1), fraction strips (BLM 9), adding machine tape cut in lengths of about 1 yard

Provide students with adding machine tape. Be sure they unroll the tape from left to right, not top to bottom.

◆ **Say:** *When you write a fraction in higher terms you multiply the numerator and the denominator by the same number. Remember: The* numerator *of a fraction is the number above the fraction bar and the* denominator *is the number below the fraction bar.*

◆ **Say:** *On the adding machine tape, draw line segments about an inch apart to form boxes from left to right. Then write 1/8 in the first box on the tape. As you multiply the numerator and denominator of 1/8 by 2, 3, 4, 5, and 6 out loud, write your answer in each box. You should have 6 boxes when you finish.*

◆ **Say:** *Tear off your fraction tape for 1/8. Read out loud each fraction in order, from left to right, on your tape.* (1/8, 2/16, 3/24, 4/32, 5/40, 6/48)

◆ **Say:** *You can use your tape to find 1/8 in higher terms by reading along the tape from left to right.*

You can have students repeat the activity for 3/5 or create a fraction tape for any fraction they choose. Be sure the fraction students start with is in simplest form.

USING PAGES IW17 and IW18

Have a student read aloud the material in the box at the top of page IW17. Relate the definition given to the fraction tape students made earlier in *Getting Started* in the following way.

◆ **Say:** *Look at your fraction tape for 1/8. What can you say about the fractions on your fraction tape.* (The fractions are equivalent fractions.)

Exercises 1–6 on page IW17

Have a student read aloud the direction line above exercises 1-2. Next have a student read aloud the first sentence above the rectangles. Have students count out loud the number of squares inside each

rectangle to verify that each rectangle is 4 squares by 6 squares.

◆ **Say:** *Now that you see that each rectangle is the same size, look at the number of parts in each. The shaded part shows the numerator while the total number of parts shows the denominator of the fraction. You see the fraction to the right of each rectangle. Remember: Each whole rectangle represents 1.*

You can have students verify this by saying out loud the number of shaded parts out of the total number of parts in each rectangle. In exercise 1 for example, 1 out of 2 parts, 3 out of 6 parts, and so on.

Finally have a student read aloud the number sentence below the rectangles. If students have difficulty expressing this statement verbally, have them count out loud the number of shaded small squares out of the total number of small squares in each rectangle. For example, 12 out of 24, 6 out of 12, and so on, for each rectangle.

In exercise 2 guide students to write the equivalent fractions shown by the rectangles in the space provided below. Point out that the fractions are written in the order of the numbers (from least to greatest) used to generate the higher terms. Stated another way, the fraction whose numerator and denominator is multiplied by 4 is written after the fraction multiplied by 2/2.

If you feel that this is too difficult for students, there is no need to stress this point. However, this patterning of multiples could help students decide whether a fraction is in simplest form.

Next have a student read aloud the direction line above exercises 3-6. Relate the fraction tapes shown to the one(s) they made earlier in *Getting Started*.

If students have difficulty with these exercises, have them stop and make fraction tapes for 2/3, 1/4, 1/3, and 5/8. Then they can use the tapes to help them answer exercises 3-6.

By analyzing equal rectangles and fraction tapes these six exercises guide students to write equivalent fractions.

Exercises 7–14 on page IW18
Have a student read aloud the material in the box at the top of page IW18. Then have a student read aloud the direction line above exercises 7-10.

◆ **Say:** *In these exercises the denominators are alike, so you only need one kind of fraction strip for each exercise. Look at exercise 7. What kind of fraction strip do you need? Why?* (You need the fraction strip for thirds because both denominators are 3.)

◆ **Say:** *You find the sum of the two fractions by counting the shaded parts of the strip for the numerators; the kind of strip — for example thirds, in exercise 7 — reflects the number in all and is the denominator.*

◆ **Say:** *You may want to shade actual fraction strips for thirds to show 1/3 and 1/3. When you place each strip one below the other you can easily see the shaded parts of the whole to find the sum.*

You can assist students as necessary to complete exercises 8-10 in the same way. Point out that when subtracting like fractions, you cross out the appropriate number of shaded parts on the strip. If students use actual strips they can cut away the appropriate number of shaded parts.

Then have a student read aloud the material in the box in the middle of page IW18 and the direction line above exercises 11-12.

◆ **Say:** *Look at exercise 11. What can you say about these denominators?* (They are unlike, or not the same.)

◆ **Say:** *When you add or subtract fractions with unlike denominators you must find a denominator that is the same, or common, for both fractions. You can write both fractions in higher terms until you find fractions with like denominators.*

In exercises 11-12 the fraction strips for the like denominators are provided because one denominator is a factor of the other.

Have a student read aloud the direction line above exercises 13–14. In these exercises students shade the strips on their own to find the sum or difference in simplest form. Assist students as necessary to write their answers in simplest form. Notice that the correct strip is provided for exercise 13. Students must subdivide the strip into sixths in exercise 14.

These eight exercises guide students to add and subtract fractions with like and unlike denominators and write their answers in simplest form.

SUMMING-IT-UP QUIZ

Give this quiz to students.

◆ **Say:** *Write the fractions 2/5, 12/20, and 9/36 and write any three equivalent fractions for each.* (sample answer: 4/10, 6/15, 8/20; 3/5, 6/10, 9/15; 1/4, 2/8, 3/12)

◆ **Say:** *Write 16/28 and 24/36. Are these fractions equivalent?* (No)

◆ **Say:** *Write these exercises. Write each sum or difference in simplest form.*
    5/16 + 7/16   1/2 + 3/8   1/2 − 3/10
(3/4; 7/8; 1/5)

If students answer the *Summing-It-Up Quiz* correctly, you may want to assign the review lessons on pages 15 and 16 of *Skills Update* in the student textbook. Then students should return to Lesson 6-3 or 6-5 and resume work on page 207 or 211 of their textbook. If students do not successfully complete the *Summing-It-Up Quiz*, further remediation may be necessary.

Answers: Pages IW17 and IW18

2. 2/10; 4/20; 8/40   **3.** 2/3; 12/18; 14/21; 16/24

4. 1/4   **5.** 1/3   **6.** 5/8

7. 2/3   **8.** 1   **9.** 1/2

10. 4/10; 2/5   **11.** 3/4   **12.** 2/10; 1/10

13. 9/12 + 2/12; 11/12   **14.** 4/6 − 3/6; 1/6

Meaning of Mixed Numbers

Use with Lesson 6-4, exercises 1–9, text pages 208–209.

DIAGNOSTIC INTERVIEW

Before beginning this *Intervention Workshop* lesson, have students complete the following exercises.

Diagnostic Exercises

◆ **Say:** *Write 25, 2.5, and 2 1/5 and tell which is a mixed number.* (2 1/5)

◆ **Say:** *Write these numbers in simplest form.*
     8/5   4 5/108 10/616/12 3 4/4
(1 3/5; 4 1/2; 9 2/3; 1 1/3; 4)

If students answer the Diagnostic Exercises correctly, there is probably no need for them to do this *Intervention Workshop* lesson. Have those students rejoin their class on page 209 of the student textbook. For students who have had difficulty answering the exercises above, continue the *Intervention Workshop*.

BACKGROUND

This *Intervention Workshop* lesson addresses the problem of the meaning of mixed numbers. Students begin with analyzing on grid paper equal rectangles that show mixed numbers. Analyzing the wholes and the number of equal parts provides a clear visual model to help students understand mixed numbers. Then students use patterns to label tick marks for missing mixed numbers on number lines. Finally they use fraction strips to write mixed numbers in simplest form.

TEACHING THE *INTERVENTION WORKSHOP* LESSON

Getting Started

Materials: fraction strips (BLM 9)

◆ **Say:** *Write these division expressions.*

     34 ÷ 6   90 ÷ 8   38 ÷ 4

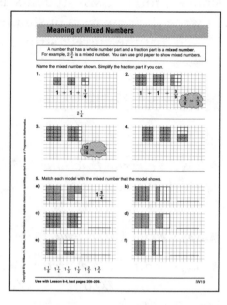

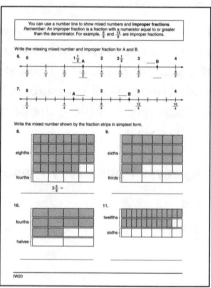

◆ **Say:** *Compute and write each quotient with a remainder as a fraction in simplest form.* (5 2/3; 11 1/4; 9 1/2)

◆ **Say:** *Write the following addition expressions. Compute and write each sum in simplest form.*

$$2/3 + 1/3 \qquad 3/6 + 5/6 + 1/6 + 3/6$$

(3/3 = 1; 12/6 = 2)

🕐 USING PAGES IW19 and IW20

Have a student read aloud the material in the box at the top of page IW19.

Relate the definition given to the quotients students computed earlier in *Getting Started* in the following way.

◆ **Say:** *Look at the quotients you wrote earlier. What can you say about them?* (The quotients are mixed numbers.)

Exercises 1–5 on page IW19
Have a student read aloud the direction line above exercises 1-4.

◆ **Say:** *Look at a large square in exercise 1. Count the number of small squares inside. How many are there? What does this tell you?* (4; This tells you the number of equal parts of the whole.)

Guide students as necessary to complete exercises 2-4. Assist students who have difficulty writing the fraction parts of the mixed numbers in simplest form.

◆ **Say:** *You can use fraction strips to help you write the fraction parts of the mixed numbers in simplest form. For example in exercise 2, show the fraction part 3/9 on the ninths strip. Then find a fraction strip that you can match up with the equal parts for ninths.*

Have a student read aloud the direction line in exercise 5. Assist students as necessary to complete this exercise. You can suggest that students use fraction strips.

By analyzing the wholes and the number of equal parts, these five exercises guide students in understanding the meaning of mixed numbers.

Exercises 6–12 on page IW20
Have a student read aloud the material in the box at the top of page IW20. Then have a student read aloud the direction line above exercises 6-7.

Relate the definition given to the sums students computed earlier in *Getting Started* in the following way.

◆ **Say:** *Look at the sums you wrote earlier. What can you say about them?* (The sums are improper fractions that were written as whole numbers.)

Guide students through exercise 6.

◆ **Say:** *Look at the numbers written above the tick marks on the number line. What kinds of numbers do you see?* (whole numbers and mixed numbers)

◆ **Say:** *Now look at the numbers written below the tick marks on the number line. What kinds of numbers do you see here?* (improper fractions)

◆ **Say:** *Look at the tick mark labeled 2. What can you say about 2 and the 4/2 written directly below it on the number line?* (The numbers are equal: 2 = 4/2.)

You can guide students to begin exercise 7 the same way. Then have a student read aloud the direction line above exercises 8–11. Assist students as necessary to write the fraction parts in simplest form. Notice that the correct strips are provided.

◆ **Say:** *You may want to use actual fraction strips to help you write the mixed numbers in simplest form. In this way you will see the whole-number part and the fraction part of each mixed number.*

These six exercises guide students to write improper fractions as mixed numbers in simplest form by using number lines and fraction strips.

SUMMING-IT-UP QUIZ

Give this quiz to students.

◆ **Say:** *Write these numbers as whole numbers or mixed numbers in simplest form.*

    14/4   16/6   20/5   2 12/12

(3 1/2; 2 2/3; 4; 3)

◆ **Say:** *Write 28/16, 12/9, and 52/32. On a number line, what whole number or mixed number would be labeled for each?* (1 3/4; 1 1/3; 1 5/8)

If students answer the *Summing-It-Up Quiz* correctly, then they should return to Lesson 6-4 and resume work on page 209 of their textbook. If students do not successfully complete the *Summing-It-Up Quiz*, further remediation may be necessary.

Answers: Pages IW19 and IW20

2. 2 1/3	**3.** 1 3/4	**4.** 2 1/2

5b. 1 1/3   **5c.** 1 3/4   **5d.** 1 1/2   **5e.** 1 1/4   **5f.** 1 1/3

6. 3 1/2; 7/2   **7.** A: 1 1/4; 5/4;   **8.** 3 3/4   **9.** 3 2/6 = 3 1/3
                              B: 2 3/4; 11/4

10. 2 2/4 = 2 1/2   **11.** 1 10/12 = 1 5/6

Operation Sense: Fractions and Mixed Numbers

Use with Lesson 6-6, exercises 17–36, text pages 212–213.

DIAGNOSTIC INTERVIEW

Before beginning this *Intervention Workshop* lesson, have students complete the following exercises.

Diagnostic Exercises

Write this exercise on the board.

$$\begin{array}{r} 8 \\ -\ 2\ 1/4 \\ \hline \end{array}$$

◆ **Say:** *Will the answer to this subtraction exercise be greater than 6 or less than 6?* (less than 6) *How do you know?* (Possible response: $8 - 2 = 6$. Since 2 1/4 is greater than 2, subtracting 2 1/4 from 8 will give you a lesser number than subtracting 2 from 8.)

◆ **Say:** *What can you do to complete the subtraction?* (Rename 8 as 7 4/4. Subtract the fractions. Subtract the whole numbers.)

Write this exercise on the board.

$$\begin{array}{r} 8\ 1/4 \\ -\ 2 \\ \hline \end{array}$$

◆ **Say:** *Will the answer to this subtraction exercise be greater than 6 or less than 6?* (greater than 6) *How do you know?* (Possible response: $8 - 2 = 6$. Since 8 1/4 is greater than 8, subtracting 2 from 8 1/4 will give you a greater number than subtracting 2 from 8.)

◆ **Say:** *What can you do to complete the subtraction?* (Subtract zero from, or bring down, the fraction. Subtract the whole numbers.)

Write this exercise on the board.

$$\begin{array}{r} 8\ 1/4 \\ -\ 2\ 3/4 \\ \hline \end{array}$$

◆ **Say:** *Will the answer to this subtraction exercise be greater than 6 or less than 6?* (less than 6) *How do you know?* (Possible response: Since 3/4 is greater than 1/4, you will need to rename 8 1/4 as 7 5/4 in order to subtract. When you subtract the ones, you will be subtracting $7 - 2$, which equals 5. The difference of the fraction parts of the mixed numbers is less than 1, so the total difference is less than 6.)

◆ **Say:** *What do you need to do to complete the subtraction?* (Rename 8 1/4 as 7 5/4. Subtract the fractions. Subtract the whole numbers.)

If students answer the Diagnostic Exercises correctly, there is probably no need for them to do this *Intervention Workshop* lesson. Have those students rejoin their class on page 213 of the student textbook. For students who have had difficulties answering the exercises above, continue the *Intervention Workshop*.

BACKGROUND

Exercises 17–36 on page 213 of the student textbook require students to subtract with mixed numbers and whole numbers. These *Intervention Workshop* exercises address the needs of students who have difficulty deciding when to rename the minuend, completing subtraction that involves renaming, and using estimation to decide when a difference is reasonable. In Activity One students estimate differences of mixed numbers and whole numbers and choose reasonable answers. In Activity Two students use models to decide when to rename minuends and to complete subtraction involving mixed numbers and whole numbers. Activity Three provides an alternate method of subtracting mixed numbers from whole numbers. Use one or more of these activities with your students, as needed.

TEACHING THE *INTERVENTION WORKSHOP* LESSON

Getting Started

Write the following on the board.

$$\begin{array}{r} 5\ 1/8 \\ -\ 3\ 3/8 \end{array}$$

◆ **Say:** *What is the fraction part of each number?* (1/8; 3/8) *Can you subtract the fraction parts without renaming?* (No) *What do you have to do in order to subtract the fraction part?* (Rename 5 1/8 as 4 9/8 so that there are enough eighths to subtract.)

ACTIVITY ONE: Use Number Sense to Check Differences for Reasonableness

Write the following on the board.

$$\begin{array}{r} 7 \\ -\ 3\ 3/5 \end{array} \qquad \text{4 3/5 or 3 2/5?}$$

◆ **Say:** *What is 7 minus 3?* (4) *Will 7 minus 3 3/5 be greater than 4 or less than 4?* (less than 4) *How do you know?* (Possible response: 7 − 3 = 4. Since 3 3/5 is greater than 3, subtracting 7 − 3 3/5 will give you a lesser number than subtracting 7 − 3.)

◆ **Say:** *Which difference is reasonable: 4 3/5 or 3 3/5?* (3 2/5) *Why?* (3 2/5 is less than 4.)

Write the following on the board.

$$\begin{array}{r} 6\ 5/8 \\ -\ 4 \\ \hline \end{array}\qquad 2\ 5/8\ or\ 1\ 3/8?$$

◆ **Say:** *What is 6 minus 4?* (2) *Will 6 5/8 minus 4 be greater than 2 or less than 2?* (greater than 2) *How do you know?* (Possible response: $6 - 4 = 2$. Since 6 5/8 is greater than 6, subtracting $6\ 5/8 - 4$ will give you a greater number than subtracting $6 - 4$.)

◆ **Say:** *Which difference is reasonable: 2 5/8 or 1 3/8?* (2 5/8) *Why?* (2 5/8 is greater than 2.)

Write the following on the board.

$$\begin{array}{r} 9\ 4/9 \\ -\ 3\ 5/9 \\ \hline \end{array}\qquad 6\ 1/9\ or\ 5\ 8/9?$$

◆ **Say:** *What is 9 minus 3?* (6) *Will 9 4/9 minus 3 5/9 be greater than 6 or less than 6?* (less than 6) *How do you know?* (Possible response: Since 5/9 is greater than 4/9, you will need to rename 9 4/9 as 8 13/9 in order to subtract. When you subtract the ones, you will be subtracting $8 - 3$. $8 - 3 = 5$, and the difference of the fractional parts is less than 1.)

◆ **Say:** *Which difference is reasonable: 6 1/9 or 5 8/9?* (5 8/9) *Why?* (5 8/9 is less than 6.)

◆ **Say:** *Suppose you changed the number being subtracted to 3 2/9. Would the difference be greater or less than 6?* (greater than 6) *Why?* (You can subtract 2/9 from 4/9 without renaming, and $4/9 - 2/9$ is greater than 0.)

Have students complete the following exercises.

- Which could be the difference of $5\ 5/8 - 2$: 3 5/8 or 2 3/8? (3 5/8)
- Which could be the difference of $8 - 4\ 3/7$: 4 3/7 or 3 4/7? (3 4/7)
- Which could be the difference of $11\ 2/3 - 6\ 1/3$: 5 1/3 or 4 1/3? (5 1/3)
- Which could be the difference of $9\ 2/9 - 2\ 4/9$: 7 2/9 or 6 7/9 (6 7/9)

ACTIVITY TWO: Use Models to Decide When to Rename Minuends

Materials: fraction strips (BLM 9)

Write the following on the board.

$$\begin{array}{r} 4\ 3/4 \\ -\ 2 \\ \hline \end{array}$$

Have students model 4 3/4 by shading 4 whole fraction strips and 3 parts on the fourths fraction strip.

◆ **Say:** *What are you subtracting from 4 3/4?* (2) *Do you need to subtract any of the fourths to complete the subtraction?* (No) *Do you need to rename a whole as fourths to complete the subtraction?* (No) *How do you use the models to show subtracting 2 from 4 3/4?* (Take away 2 whole fraction strips.) *Take away two whole fraction strips. What mixed number is shown by the models now?* (2 3/4)

Write the following on the board.

$$4$$
$$- \ 1 \ 5/8$$

Have students model 4 by shading 4 whole fraction strips.

◆ **Say:** *What are you subtracting from 4?* (1 5/8) *Are there any eighths to subtract 5/8 from?* (No) *How can you get eighths so that you can subtract 5/8?* (Trade 1 whole fraction strip for 1 eighths fraction strip.)

◆ **Say:** *Do you need to rename a whole as eighths to complete the subtraction?* (Yes)

Have students trade 1 whole fraction strip for 1 eighths fraction strip and shade the entire eighths strip. Point out that the 3 whole fraction strips and the eighths fraction strip show the minuend, 4, which students renamed as 3 and 8/8. Have students cover 1 whole and 5/8 with unshaded fraction strips to show the subtraction.

◆ **Say:** *What mixed number do the shaded parts show now?* (2 3/8) *What is 4 minus 1 5/8?* (2 3/8)

Write the following on the board.

$$5 \ 1/4$$
$$- \ 3 \ 3/4$$

Have students model 5 1/4 by shading 5 whole fraction strips and 1 part on the fourths fraction strip.

◆ **Say:** *What are you subtracting from 5 1/4?* (3 3/4) *Are there enough fourths to subtract 3/4 from?* (No) *How can you get some more fourths so that you can*

subtract 3/4? (Trade 1 whole fraction strip for 1 fourths fraction strip.)

◆ **Say:** *Do you need to rename a whole as fourths to complete the subtraction?* (Yes)

Have students trade 1 whole fraction strip for 1 fourths fraction strip and shade the entire fourths strip. Point out that students now have four whole fraction strips, a fraction strip that shows 4/4, and a fraction strip that shows 1/4. These fraction strips show the minuend, 5 1/4, which students have renamed as 4 5/4. Have students cover 3 wholes and 3/4 with unshaded fraction strips to show the subtraction.

◆ **Say:** *What mixed number do the shaded parts show now?* (1 2/4) *What is 1 2/4 in simplest form?* (1 1/2) *What is 5 1/4 minus 3 3/4?* (1 1/2)

Write the following on the board.

$$3 \ 4/5$$
$$- \ 2 \ 3/5$$

Have students model 3 4/5 by shading 3 whole fraction strips and 4 parts on the fifths fraction strip.

◆ **Say:** *What are you subtracting from 3 4/5?* (2 3/5) *Are there enough fifths to subtract 3/5 from?* (Yes) *Do you need to rename a whole as fifths to complete the subtraction?* (No)

Have students cover 2 wholes and 3/5 with unshaded fraction strips to show the subtraction.

◆ **Say:** *What mixed number do the shaded parts show now?* (1 1/5) *What is 3 4/5 minus 2 3/5?* (1 1/5)

ACTIVITY THREE: Use Operation Sense to Subtract Mixed Numbers and Whole Numbers

Write the following on the board.

$$9 - 3\ 2/5$$

◆ **Say:** *Let's subtract the whole number part first. What is 9 minus 3?* (6) *What do we still have to subtract?* (2/5) *Are there any fifths to subtract 2/5 from?* (No) *How can you get fifths so that you can subtract 2/5?* (Rename 1 of the 6 ones as 5/5.) *Rename a 1 as 5/5. What mixed number does that give you?* (5 5/5) *Subtract 2/5 from 5/5. How many fifths are left?* (3/5) *How many ones are left?* (5) *What is 9 minus 3 2/5?* (5 3/5)

Have students use the above method independently to find other differences involving whole numbers and mixed numbers such as 5 − 2 7/8 and 10 − 4 5/9. (2 1/8; 5 4/9)

SUMMING-IT-UP QUIZ

Give this quiz to students.

Write the following on the board.

$$11 - 3\ 4/7$$
$$7\ 8/9 - 5$$
$$6\ 1/5 - 2\ 3/5$$
$$10\ 5/9 - 4\ 4/9$$

◆ **Say:** *Which could be the difference of 11 minus 3 4/7: 8 4/7 or 7 3/7?* (7 3/7) *Which could be the difference of 7 8/9 minus 5: 2 8/9 or 1 1/9?* (2 8/9) *Which could be the difference of 6 1/5 minus 2 3/5: 4 2/5 or 3 3/5?* (3 3/5) *Which could be the difference of 10 5/9 minus 4 4/9: 6 1/9 or 5 1/9?* (6 1/9)

Write the following on the board.

$$3 - 2\ 3/5$$
$$5\ 3/5 - 3$$
$$4\ 4/5 - 1\ 2/5$$
$$9\ 2/5 - 2\ 3/5$$

◆ **Say:** *In which exercises do you have to rename in the first number in order to subtract?* (3 − 2 3/5; 9 2/5 − 2 3/5)

Have students complete the two subtraction exercises they named above [(3 − 2 3/5; 9 2/5 − 2 3/5)], using fraction strips if they wish. (2/5; 6 4/5)

If students answer the *Summing-It-Up Quiz* correctly, they should then resume work on page 213 of their textbook. If students do not successfully complete the *Summing-It-Up Quiz,* further remediation may be necessary.

Use with Lesson 7-1, exercises 3–17, text pages 226–227.

DIAGNOSTIC INTERVIEW

Before beginning this *Intervention Workshop* lesson, have students complete the following exercises.

Diagnostic Exercises

◆ **Say:** *Write the following multiplication expressions, multiply, and then write the product in simplest form.*

   $3/4 \times 4/6$ $3/8 \times 4/12$ $3/10 \times 5/9$ $5/8 \times 6/10$

(1/2; 1/8; 1/6; 3/8)

◆ **Say:** *Write the following pattern. Continue the pattern, then write the rule you discovered.*

$1/5 \times 1/2 = 1/10$	$1/5 \times 1/6 =$
$1/5 \times 1/3 = 1/15$	$1/5 \times 1/7 =$
$1/5 \times 1/4 =$	$1/5 \times 1/8 =$
$1/5 \times 1/5 =$	$1/5 \times 1/9 =$

(1/20; 1/25; 1/30; 1/35; 1/40; 1/45; Multiply the numerators and multiply the denominators.)

If students answer the Diagnostic Exercises correctly, there is probably no need for them to do this *Intervention Workshop* lesson. Have those students rejoin their class on page 227 of the student textbook. For students who have had difficulty answering the exercises above, continue the *Intervention Workshop*.

BACKGROUND

This *Intervention Workshop* lesson addresses the problem of finding common factors to cancel when multiplying fractions. Students begin by multiplying fractions with no common factors, using rectangles. Counting the shaded squares inside the rectangle provides a clear visual methodology for students to readily determine the product. Follow-up is provided by multiplying fractions with no common factors, using patterns. Students then multiply fractions with common factors, using this same methodology. At the end of the lesson students again complete patterns of multiplying fractions with common factors.

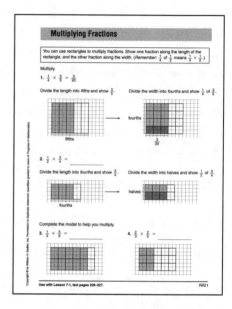

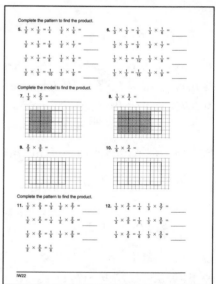

TEACHING THE *INTERVENTION WORKSHOP* LESSON

Getting Started

Materials: counters

Provide students with counters.

◆ **Say:** *You can use counters to find the answer to 2/3 of 12. Separate the 12 counters into 3 equal groups. The number of counters in two of the equal groups is 2/3 of 12.*

Draw this diagram for students.

◆ **Say:** *What is 2/3 of 12?* (8)

◆ **Say:** *Write the exercises 3/5 of 20, 1/3 of 15,* and *3/4 of 12. Use counters to help you find each answer.* (12; 5; 9)

◐ USING PAGES IW21 and IW22

Have a student read aloud the material in the box at the top of page IW21. Relate the *Remember* note to the exercises students computed in *Getting Started* in the following way.

◆ **Say:** *Look at the exercises you wrote earlier. How do they relate to the Remember statement?* (3/5 of 20 is the same as 3/5 × 20; 1/3 of 15 is the same as 1/3 × 15; 3/4 of 12 is the same as 3/4 × 12.)

Exercises 1–4 on page IW21
Have a student read aloud the direction line above exercises 1–2.

◆ **Say:** *Look at the first rectangle in exercise 1. Notice that the length is divided into fifths. Three parts are shaded to show 3/5. In the second rectangle the width is divided into fourths. One part is shaded to show 1/4. The parts that are shaded twice show the answer. You count these to find the answer, 3/20.*

◆ **Say:** *If you count the total number of small squares inside the second rectangle you will see that 3 out of the 20 are shaded twice. So you can see that 3/20 is the product.*

You can guide students to complete exercise 2 the same way. Have a student read aloud the direction line above exercises 3 and 4. Assist students as necessary to complete these exercises.

By analyzing the number of equal parts shaded twice in rectangles, exercises 1–4 guide students in multiplying fractions with no common factors.

Exercises 5–6 on page IW22
Have a student read aloud the direction line above exercises 5–6.

◆ **Say:** *Notice that the numerator in the product is the answer when you multiply the two numerators. Similarly the denominator in the product is the answer when you multiply the two denominators.*

The patterns in exercises 5–6 give students further guidance in multiplying fractions with no common factors.

Exercises 7–12 on page IW22
Have a student read aloud the direction line above exercises 7–10 on page IW22.

◆ **Say:** *These exercises are like the ones you completed on page 21.*

Assist students as necessary to complete these exercises. Then have a student read aloud the direction line above exercises 11–12. Guide students to complete the patterns.

◆ **Say:** *Notice that the numerator in the product is the same as the numerator in the first fraction and that the denominator in the product is the same as the denominator in the second fraction. Also notice in exercise 11 that 2 is always the denominator of the first fraction and the numerator of the second fraction. In exercise 12, 3 is always the denominator of the first fraction and the numerator of the second fraction. The 2 and 3 are called common factors. Common factors in the numerator and denominator need not be multiplied because they are equivalent to 1; that is, 2/2 = 1, 3/3 = 1.*

Write the first product in exercise 11 to show what you explained.

$$\frac{1}{\cancel{2}_1} \times \frac{\cancel{2}^1}{3} = \frac{1}{1} \times \frac{1}{3} = \frac{1}{3}$$

These four exercises (7–10) and the two follow-up pattern exercises (11–12) guide students to multiply fractions with common factors.

SUMMING-IT-UP QUIZ

Give this quiz to students.

◆ **Say:** *Write the multiplication expressions 1/2 × 3/5, 4/5 × 5/8, and 3/10 × 5/6. Multiply and write the product in simplest form.* (3/10; 1/2; 1/4)

◆ **Say:** *Copy and complete this pattern. Tell the common factor.*

1/4 × 4/5 = 1/5	1/4 × 4/8 = (1/8)
1/4 × 4/6 = (1/6)	1/4 × 4/9 = (1/9)
1/4 × 4/7 = (1/7)	(The common factor is 4.)

If students answer the *Summing-It-Up Quiz* correctly, then they should return to Lesson 7-1 and resume work on page 227 of their textbook. If students do not successfully complete the *Summing-It-Up Quiz,* further remediation may be necessary.

Answers: Pages IW21 and IW22

2. 3/8   **3.** 5/12   **4.** 4/15

5. 1/12; 1/14; 1/16; 1/18   **6.** 1/18; 1/21; 1/24; 1/27   **7.** 2/6 or 1/3

8. 3/12 or 1/4   **9.** 6/15 or 2/5   **10.** 3/24 or 1/8

11. 1/7; 1/8; 1/9   **12.** 1/7; 1/8; 1/9

Multiply Fractions by Fractions or Whole Numbers

Use with Lesson 7-1, exercises 3–17, text pages 226–227 and Lesson 7-2, exercises 1–24, text pages 228–229.

DIAGNOSTIC INTERVIEW

Before beginning this *Intervention Workshop* lesson, have students complete the following exercises.

Diagnostic Exercises

Write this exercise on the board.

> 2/3 × 1/4

◆ **Say:** *Will this product be greater than 1/4 or less than 1/4?*
(less than 1/4) *How do you know?* (Possible responses: 1 × 1/4 = 1/4. 2/3 is less than 1. So 2/3 × 1/4 will have a product that is less than 1/4; when you multiply 2/3 × 1/4 you are finding a part of 1/4. So the product will be less than 1/4.)

Write this exercise on the board.

> 3 × 2/3

◆ **Say:** *Will this product be greater than 2/3 or less than 2/3?*
(greater than 2/3) *How do you know?* (Possible response 3 × 2/3 means three 2/3s. So 3 × 2/3 will have a product that is greater than 2/3.)

◆ **Say:** *Will this product be greater than 3 or less than 3?* (less than 3) *How do you know?* (Possible response: 3 × 1 = 3. 2/3 is less than 1. So 3 × 2/3 will have a product that is less than 3.)

If students answer the Diagnostic Exercises correctly, there is probably no need for them to do this *Intervention Workshop* lesson. Have those students rejoin their class on page 227 or 229 of the student textbook. For students who have had difficulties answering the exercises above, continue the *Intervention Workshop*.

BACKGROUND

Exercises 3–17 on page 227 of the student textbook require students to multiply fractions by fractions. Exercises 1–24 on page 229 of the student textbook require students to multiply fractions and whole numbers. These *Intervention Workshop* activities address the needs of students who have difficulty predicting the magnitude of products involving fractions and using these predictions to decide if a product is reasonable. In Activity One students use models to derive a rule for predicting the magnitude of a product when a fraction is multiplied by a fraction. In Activity Two students use models to derive a rule for predicting the magnitude of a product when a fraction and a whole number are multiplied. In Activity Three students use compatible numbers to estimate the product when a fraction and whole number are multiplied. Use one or more of these activities with your students, as needed.

TEACHING THE *INTERVENTION WORKSHOP* LESSON

Getting Started

Write the following on the board.

> 2 × 3/4   2/5 × 3/4

◆ **Say:** *How are the problems alike?* (In each problem 3/4 is being multiplied by a number.) *How are the problems different?* (In the first problem, 3/4 is being multiplied by 2. In the second, 3/4 is being multiplied by 2/5.) *Which product do you think is greater?* (2 × 3/4) *Why?* (possible response: 2 is greater than 2/5, so 2 × 3/4 will be greater than 2/5 × 3/4.)

ACTIVITY ONE: Use Models to Predict the Magnitude of a Product of Two Fractions

Materials: fraction strips (BLM 9)

Write the following on the board.

> 1/2 × 3/4
>
> 3/4 × 1/2

Have students show 3/4 by shading 3 parts of a fraction strip for fourths.

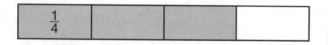

◆ **Say:** *Suppose you cut the shaded part of your fraction strip in half. Would the part of the strip you cut off be greater or less than 3/4 of the whole strip?* (less than 3/4) *How do you know?* (Possible response: The shaded part is 3/4 of the strip. Half of the shaded part is less than 3/4.)

◆ **Say:** *Is 1/4 times 3/4 greater or less than 3/4?* (less than 3/4)

Now have students show 1/2 by shading 1 part of a fraction strip for halves.

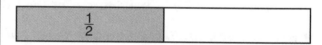

◆ **Say:** *Suppose you folded the shaded part of your fraction strip into fourths, and cut off three of them . Would the part of the strip you cut off be greater or less than half of the whole strip?* (less than half) *How do you know?* (Possible response: The shaded part is 1/2 of the strip. 3/4 of the shaded part is less than 1/2.)

◆ **Say:** *Is 3/4 times 1/2 greater or less than 1/2?* (less than 1/2)

Write the following on the board.

> 1/3 × 3/5
>
> 3/5 × 1/3

Have students work in pairs to estimate whether the product of 1/3 × 3/5 will be greater or less than 3/5, and whether the product of 3/5 × 1/3 will be greater or less than 1/3. Let them use fraction strips if they wish.

◆ **Say:** *When you multiply 1/3 and 3/5, will the product be greater or less than 1/3?* (less than 1/3) *Will the product be greater or less than 3/5?* (less than 3/5) *When you multiply two fractions, will the product be greater or less than the first factor?* (less) *Will the product be greater or less than the second factor?* (less)

Have students write a rule in their own words for how the size of a product compares to the size of the factors when two fractions are multiplied.

ACTIVITY TWO: Use Models to Predict the Magnitude of a Product of a Fraction and a Whole Number

Materials: fraction strips (BLM 9)

Write the following on the board.

3 × 2/3

Have students show 3 × 2/3 by using 3 fraction strips for thirds and shading 2 parts of each one.

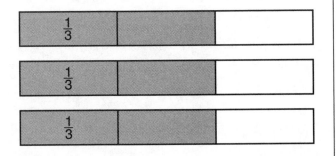

◆ **Say:** *Is 3 times 2/3 greater or less than 2/3?* (greater than 2/3) *How do you know?* (Possible response: The shaded part of each fraction strip represents 2/3. So, together, the shaded parts of 3 strips represent more than 2/3.) *Is 3 times 2/3 greater or less than 3?* (less than 3) *How do you know?* (Possible response: Together, the strips represent 3. Not all of the thirds on the strips are shaded. So 3 × 2/3 is less than 3.)

Write the following on the board.

4 × 2/5

Have students work in pairs to estimate whether the product of 4 × 2/5 will be greater or less than 4, and whether it will be greater or less than 2/5. Let them use fraction strips if they wish.

◆ **Say:** *When you multiply 4 times 2/5, will the product be greater or less than 4?* (less than 4) *Will the product be greater or less than 2/5?* (greater than 2/5) *When you multiply a fraction and a whole number, will the product be greater or less than the fraction factor?* (greater) *Will the product be greater or less than the whole-number factor?* (less)

Have students write a rule in their own words for how the size of a product compares to the size of the factors when one factor is a fraction and one factor is a whole number.

ACTIVITY THREE: Use Compatible Numbers to Predict the Magnitude of the Product of a Fraction and a Whole Number

Write the following on the board.

$$1/5 \times 31 \qquad 31 \div 5$$

Lead students through the process for using compatible numbers to estimate the product when the factors are a fraction and a whole number.

◆ **Say:** *How is multiplying a number by 1/5 like dividing a number by 5?* (Possible response: When you multiply a number by 1/5, you are finding 1 of 5 equal parts of a number. When you divide a number by 5, you are finding 1 of 5 equal parts of a number.)

◆ **Say:** *When you use compatible numbers to estimate a quotient, you find a number close to the dividend that can be divided evenly by the divisor. What compatible numbers would you use to estimate 31 divided by 5?* (30 ÷ 5) *What is 30 divided by 5?* (6) *About how much is 31 divided by 5?* (6) *About how much is 1/5 times 31?* (6)

Write the following on the board.

$$1/4 \times 19$$
$$3/4 \times 19$$

◆ **Say:** *What division example has the same meaning as 1/4 times 19?* (19 ÷ 4) *What compatible numbers would you use to estimate 19 divided by 4?* (20 ÷ 4) *How much is 20 divided by 4?* (5) *About how much is 19 divided by 4?* (5) *About how much is 1/4 times 19?* (5) *How could you use the estimate you just made for 1/4 times 19 to help you estimate 3/4 times 19?* (Multiply the estimate you just made by 3.) *About how much is 3/4 times 19?* (about 3 × 5, or 15)

Have students work independently to estimate each of the following pairs of products.

1/9 × 26 (about 3)	4/9 × 26 (about 12)
1/6 × 38 (about 6)	5/6 × 38 (about 30)
1/10 × 69 (about 7)	3/10 × 69 (about 21)

SUMMING-IT-UP QUIZ

Give this quiz to students.

Write the following on the board.

$$2/5 \times 5/8 \qquad 1/4, 1/2, \text{ or } 1\ 1/4?$$

◆ **Say:** *Is the product greater or less than 2/5?* (less than 2/5) *Is the product greater or less than 5/8?* (less than 5/8) *Which could be the product: 1/4, 1/2, or 1 1/4?* (1/4)

Write the following on the board.

$$4 \times 5/6 \qquad 5/24, 3\ 1/3, \text{ or } 5\ 1/6$$

◆ **Say:** *Is the product greater or less than 4?* (less than 4) *Is the product greater or less than 5/6?* (greater than 5/6) *Which could be the product: 5/24, 3 1/3, or 5 1/6?* (3 1/3)

If students answer the *Summing-It-Up Quiz* correctly, they should then resume work on page 227 or 229 of their textbook. If students do not successfully complete the *Summing-It-Up Quiz,* further remediation may be necessary.

Fraction Form with Operations

Use with Lesson 7-4, exercises 9–31 text pages 232–233 and Lesson 7-9, exercises 3–34, text pages 242–243.

DIAGNOSTIC INTERVIEW

Before beginning this *Intervention Workshop* lesson, have students complete the following exercises.

Diagnostic Exercises

Write this exercise on the board.

  3 1/6 × 3 3/8

◆ **Say:** *What do you need to do to each factor before you can do the multiplication?* (Rename each factor as an improper fraction.)

Have students rename each factor as an improper fraction.

◆ **Say:** *What are the improper fractions?* (19/6 and 27/8)

Write this exercise on the board.

  7 3/8 ÷ 2 1/2

◆ **Say:** *What is the first step you would take in order to do this division problem?* (Rename the dividend and the divisor as improper fractions.)

Have students rename the dividend and divisor as improper fractions.

◆ **Say:** *What are the improper fractions?* (59/8 and 5/2)

If students answer the Diagnostic Exercises correctly, there is probably no need for them to do this *Intervention Workshop* lesson. Have those students rejoin their class on page 233 or 243 of the student textbook. For students who have had difficulties answering the exercises above, continue the *Intervention Workshop*.

BACKGROUND

Exercises 9–31 on page 233 of the student textbook require students to multiply with mixed numbers. Exercises 3–34 on page 243 of the student textbook require students to divide with mixed numbers. These *Intervention Workshop* activities address the needs of students who have not yet mastered the process for renaming mixed numbers as improper fractions so that the numbers can be multiplied or divided. In Activity One students use models to rename mixed numbers as fractions. In Activity Two students review a computational process for renaming mixed numbers as fractions. Activity Three provides an opportunity for students to practice renaming mixed numbers as fractions. Use one or more of these activities with your students, as needed.

TEACHING THE *INTERVENTION WORKSHOP* LESSON

Getting Started

Write the following on the board.

     3 1/2   3/8   10/9

◆ **Say:** *Which number is an improper fraction?* (10/9) *How do you know?* (Possible responses: Its numerator is greater than its denominator; it has a value greater than 1.) *Which number is a mixed number?* (3 1/2)

Have students give examples of improper fractions and mixed numbers.

ACTIVITY ONE: Use Models to Rename Mixed Numbers as Improper Fractions

Materials: fraction strips (BLM 9)

Write the following on the board.

     2 5/8 × 1 3/4

◆ **Say:** *What do you need to do to each factor before you can do the multiplication?* (Rename each factor as an improper fraction.)

Tell students they will now use fraction strips to rewrite each factor in the multiplication problem as an improper fraction. Have students show 2 5/8 by shading 2 entire fraction strips for eighths and 5 parts of a fraction strip for eighths.

◆ **Say:** *How many eighths are shaded in all?* (21) *How many eighths are in 2 5/8?* (21) *What improper fraction can you write for 2 5/8?* (21/8)

On the board, write 21/8 under 2 5/8 as shown.

     2 5/8 × 1 3/4
       ↓

     21/8

Have students show 1 3/4 by shading 1 entire fraction strip for fourths and 3 parts of a fraction strip for fourths.

◆ **Say:** *How many fourths are shaded in all?* (7) *How many fourths are in 1 3/4?* (7) *What improper fraction can you write for 1 3/4?* (7/4)

On the board, write 7/4 under 1 3/4 as shown.

     2 5/8 × 1 3/4
       ↓   ↓

     21/8 × 7/4

◆ **Say:** *Why will multiplying 21/8 times 7/4 give you the answer to 2 5/8 times 1 3/4?* (2 5/8 = 21/8; 1 3/4 = 7/4)

Write the following on the board.

     4 3/5 ÷ 5 2/3

◆ **Say:** *What do you need to do to the dividend and the divisor before you can do the division?* (You need to rename each number as an improper fraction.) *What fraction strips could you use to help you* find an improper fraction for 4 3/5? (fraction strips for fifths) *Why would you use fraction strips for fifths?* (The fraction part of 4 3/5 is in fifths, so the improper fraction will be in fifths.)

Have students show 4 3/5 by shading fraction strips for fifths.

◆ **Say:** *How many fifths are shaded in all?* (23) *How many fifths are in 4 3/5?* (23) *What improper fraction can you write for 4 3/5?* (23/5)

On the board, write 23/5 under 4 3/5 as shown.

$$4\ 3/5 \div 5\ 2/3$$
$$\downarrow$$
$$23/5$$

◆ **Say:** *What fraction strips could you use to help you find an improper fraction for 5 2/3?* (fraction strips for thirds) *Why would you use fraction strips for thirds?* (The fraction part of 5 2/3 is in thirds, so the improper fraction will be in thirds.)

Have students show 5 2/3 by shading fraction strips for thirds.

◆ **Say:** *How many thirds are shaded in all?* (17) *How many thirds are in 5 2/3?* (17) *What improper fraction can you write for 5 2/3?* (17/3)

On the board, write 17/3 under 5 2/3 as shown.

$$4\ 3/5 \div 5\ 2/3$$
$$\downarrow \qquad\qquad \downarrow$$
$$23/5 \div 17/3$$

◆ **Say:** *Why will dividing 23/5 by 17/3 give you the answer to 4 3/5 divided by 5 2/3?* (4 3/5 = 23/5; 5 2/3 = 17/3)

Have students use fraction strips to rename mixed numbers as improper fractions in other exercises, such as 1 7/8 × 5 1/4 and 8 5/6 ÷ 3 4/9. (15/8 × 21/4; 53/6 ÷ 31/9)

ACTIVITY TWO: Use Computation to Rename Mixed Numbers as Improper Fractions

Write the following on the board.

$$4\ 1/3 \times 6\ 3/4$$

◆ **Say:** *What do you need to do to each factor before you can do the multiplication?* (Rename each factor as an improper fraction.)

Lead students through the process of renaming 4 1/3 as an improper fraction.

◆ **Say:** *How many thirds are in one whole?* (3) *How many thirds are in 2?* (6) *How many thirds are in 3?* (9) *How many thirds are in 4?* (12) *How could you use multiplication to find how many thirds*

are in 4? (Multiply 4 × 3, because there are three thirds in each whole.) *How many thirds are in 4 1/3?* (13) *How do you know?* (Possible response: There are 12 thirds in 4; so there are 12 thirds + 1 third in 4 1/3.)

◆ **Say:** *How can you use multiplication, then addition, to find the number of thirds in 4 1/3?* (Multiply the denominator of the fraction part, 3, by the whole number, 4. Add the product, 12, to the numerator 1. Write the sum over the denominator, 3.)

◆ **Say:** *How can you use multiplication, then addition, to find an improper fraction for any mixed number?* (Multiply the denominator of the fraction part by the whole number. Add the product to the numerator. Write the sum over the denominator.)

Have students rename 6 3/4 as an improper fraction. Then have them explain how they did it.

◆ **Say:** *What improper fraction did you write for 6 3/4?* (27/4) *How can you use multiplication, then addition, to find the number of fourths in 6 3/4?* (Multiply the denominator of the fraction part, 4, by the whole number, 6. Add the product, 24, to the numerator, 3.)

Have students use computation to rename the fractions as improper fractions in other exercises, such as 9 2/5 × 2 3/10 and 8 3/8 ÷ 2 1/3. (47/5 × 23/10; 67/8 ÷ 7/3)

ACTIVITY THREE: Practice Writing Improper Fractions for Mixed Numbers

Materials: fraction strips (BLM 9)

Draw the following diagram on the board.

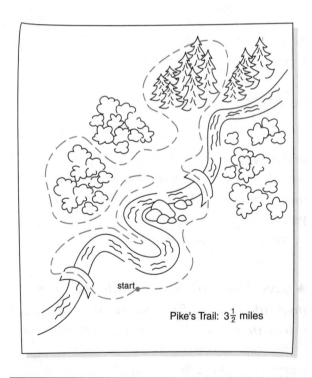

start

Pike's Trail: $3\frac{1}{2}$ miles

◆ **Say:** *Marisa runs 2 2/3 loops around Pike's Trail. What multiplication example would tell you how many miles Marisa runs in all?* (2 2/3 × 3 1/2) *What do you need to do to each factor before you can do the multiplication?* (Rename each factor as an improper fraction.)

Have students rename each factor as an improper fraction, using fraction strips if they wish.

◆ **Say:** *What improper fractions would you multiply to tell you how many miles Marisa runs in all?* (8/3 and 7/2)

Have students work in pairs. Each student writes a word problem that requires multiplication or division of mixed numbers. Students exchange problems, tell what mixed numbers they need to multiply or divide to solve the problem, then rename each mixed number as an improper fraction.

SUMMING-IT-UP QUIZ

Give this quiz to students.

Write the following on the board.

$1\ 3/5 \times 2\ 1/3$

◆ **Say:** *What do you need to do to each factor before you can do the multiplication?* (Rename each factor as an improper fraction.)

Have students rename each factor as an improper fraction, then use the improper fractions to rewrite the multiplication example.

◆ **Say:** *What multiplication sentence did you write?* ($8/5 \times 7/3$) *Why will multiplying 8/5 times 7/3 give you the answer to 1 3/5 times 2 1/3?* ($1\ 3/5 = 8/5$; $2\ 1/3 = 7/3$)

Write the following on the board.

$4\ 6/7 \div 3\ 1/6$

◆ **Say:** *What do you need to do to the dividend and the divisor before you can do the division?* (You need to rename each number as an improper fraction.)

Have students rename each number as an improper fraction, then use the improper fractions to rewrite the division example.

◆ **Say:** *What division sentence did you write?* ($34/7 \div 19/6$) *Why will dividing 34/7 by 19/6 give you the answer to 4 6/7 divided by 3 1/6?* ($4\ 6/7 = 34/7$; $3\ 1/6 = 19/4$)

If students answer the *Summing-It-Up Quiz* correctly, they should then resume work on page 233 or 243 of their textbook. If students do not successfully complete the *Summing-It-Up Quiz,* further remediation may be necessary.

Take a Survey

Use with Lesson 8-2, exercises 8–12, text pages 260–261.

DIAGNOSTIC INTERVIEW

Before beginning this *Intervention Workshop* lesson, have students complete the following exercises.

Diagnostic Exercises

Display this graph on a transparency on the overhead, or draw it on the board. Explain that the graph shows the results of a survey about favorite types of magazines among all people in a city.

◆ **Say:** *Suppose you ran a newspaper stand. Based on the results of this survey, which kind of magazine would you expect to sell the most?* (sports) *Why?* (Sports magazines got the largest number of votes.) *How could you use the survey to predict the number of sports magazines you would sell if you sold 100 magazines in all?* (Find the fractional part of those surveyed that chose sports magazines as their favorite. Multiply 100 by that fraction.)

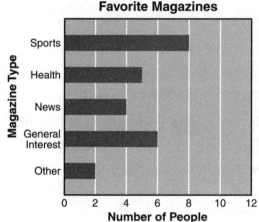

◆ **Say:** *Suppose you were going to take a survey of favorite kinds of magazines among all people in a city, like the survey in the graph. Which would be a better place to take the survey: a street corner or a sports arena?* (a street corner) *Why?* (Possible response: The street corner would give a good mixture of different kinds of people, with opinions likely to be unbiased; the sports arena would be likely to give you a sample with more sports fans than may be in the general population.)

If students answer the Diagnostic Exercises correctly, there is probably no need for them to do this *Intervention Workshop* lesson. Have those students rejoin their class on page 261 of the student textbook. For students who have had difficulties answering the exercises above, continue the *Intervention Workshop*.

BACKGROUND

Exercises 8–12 on page 261 of the student textbook require students to interpret the results of a survey, to use the results of a survey to make predictions, and to identify factors that may prevent a sample from being representative. These *Intervention Workshop* activities address the needs of students who have difficulty interpreting and using survey results and who do not yet understand the concept of a representative sample. In Activity One students interpret survey results that are presented in a bar graph. In Activity Two students use survey results from a pictograph to draw conclusions and make predictions. Activity Three provides an opportunity for students to decide whether or not a sample is representative of a given population. Use one or more of these activities with your students, as needed.

TEACHING THE *INTERVENTION WORKSHOP* LESSON

Getting Started

Display this graph on a transparency on the overhead or draw it on the board. Explain that the graph shows the results of a survey.

◆ **Say:** *What was the survey about?* (favorite sandwiches) *What does each bar on the graph show?* (the number of people who chose a particular kind of sandwich as their favorite)

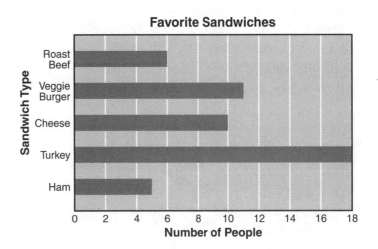

ACTIVITY ONE: Interpret Survey Results

Display the graph titled *Favorite Sandwiches* that you created for *Getting Started.* Remind students that the graph shows the results of a survey about favorite sandwiches.

◆ **Say:** *Which kind of sandwich was least popular?* (ham) *How do you know?* (The bar for ham is the shortest bar.) *Which kind of sandwich was most popular?* (turkey) *How do you know?* (The bar for turkey is the longest bar.)

◆ **Say:** *How could you find the number of people who chose turkey sandwiches as their favorite sandwich?* (Look at the bar for turkey sandwich, see where it ends, and note the number on the bottom scale.) *How many people chose turkey sandwiches?* (18)

Tell students they will now find a fraction that describes the fractional part of the people surveyed who chose turkey sandwiches.

◆ **Say:** *How could you find the total number of people who were surveyed?* (Find the number of people who chose each kind of sandwich. Add the numbers.) *What is the total number of people surveyed?* (50) *How many of those people chose turkey as their favorite sandwich?* (18) *What fraction could you write to represent 18 out of 50?* (18/50, or 9/25) *What fraction of those surveyed chose turkey sandwiches as their favorite sandwich?* (18/50, or 9/25)

Now have students write a fraction to show the fractional part of those surveyed that chose each of the other kinds of sandwiches.

◆ **Say:** *What fraction of those surveyed chose roast beef as their favorite sandwich?* (6/50, or 3/25) *What fraction chose veggie burgers?* (11/50) *What fraction chose cheese sandwiches?* (10/50, or 1/5) *What fraction chose ham sandwiches?* (5/50, or 1/10)

ACTIVITY TWO: Use Surveys to Make Predictions

Display this graph on a transparency on the overhead or draw it on the board. Explain that the graph shows the results of a survey that asks, "Do you think students should wear school uniforms? Answer Yes, No, or Not Sure."

School Uniforms					
Yes	👕	👕	👕	👕	👕
No	👕	👕	👕	👕	
Not Sure	👕	👕			
Key: Each 👕 = 4 votes.					

Explain also that the question was asked of a small part of the student population that served as a sample of the school population.

◆ **Say:** *From the survey would you predict that the greatest number of students in the school (a) want to have school uniforms, (b) do not want to have school uniforms, or (c) are not sure about whether they want school uniforms?* (*a*, want school uniforms.) *Why?* (More students surveyed answered *Yes* than *No* or *Not Sure*.)

Tell students that they will now use the data from the survey to predict the total number of students in the school who want school uniforms. First have students use the graph to find the number of students in the survey, and the number of students who said *Yes* to the survey question.

◆ **Say:** *How many students said* Yes *to the survey question?* (20) *How many students were in the survey?* (40) *What fractional part of those surveyed said* Yes *to the question?* (20/40, or 1/2)

◆ **Say:** *The school has 600 students. How could you use multiplication to predict the total number of students who want school uniforms?* (Multiply the total number of students in school, 600, by the fractional part of those surveyed who said *Yes*, 1/2.)

Have students complete the computation.

◆ **Say:** *Based on the results of the survey, how many students in the entire school population would want school uniforms?* (300 students)

ACTIVITY THREE: Decide Whether a Sample is Representative

Draw the following diagram on the board.

Population	Voters in Midwood City	Sample
The group you are getting information about	Voters chosen at random in a department store Voters chosen at random at a baseball game	The part of the population used in a survey

◆ **Say:** *Suppose you want to take a survey to find out whether the voters of Midwood City want to build a new baseball stadium. What is the population you want information from?* (the voters in Midwood City) *To take the survey you want a sample that represents a good mix of people in the population and is not biased. Which sample would you chose: voters in a department store or voters at a baseball game?* (voters in a department store) *Why?* (Possible response: Voters in a department store would give a good mixture of people with opinions more likely to be unbiased. Voters at a baseball game are more likely to have biased opinions, since the survey question is about baseball.)

For each of the following survey topics, have students write a sentence or two telling why they think the sample is or is not a good sample to use.

Survey Topic	Sample
Favorite food among all people in town	Students in the school cafeteria
Favorite class among all students in school	Members of the school's computer club
Favorite sport among school students	Students on one school bus
Favorite month of the year	Members of a skiing club
Favorite computer program among students in sixth grade	Sixth-grade students who have last names beginning with *A* to *J*

SUMMING-IT-UP QUIZ

Give this quiz to students.

Display this graph on a transparency on the overhead or draw it on the board. Explain that the graph shows the results of a survey about favorite fruits.

Favorite Fruits of Sixth Graders

◆ **Say:** *How many people in all were surveyed?* (30) *What part of the people surveyed chose oranges as their favorite fruit?* (6/30, or 1/5) *Suppose you were having a picnic lunch for an entire sixth grade of 90 students. Based on the results of the survey, how would you decide how many oranges to buy?* (Multiply 1/5 × 90) *How many oranges would you buy?* (18)

◆ **Say:** *Suppose you want to conduct a survey to find out about favorite pets. Which place would be a better place to take the survey: a dog show or a pet store?* (a pet store) *Why?* (Possible response: People chosen at random in a pet store would give a mix of people who liked different kinds of pets. People chosen at a dog show are more likely to have a bias in favor of dogs.)

If students answer the *Summing-It-Up Quiz* correctly, they should then resume work on page 261 of their textbook. If students do not successfully complete the *Summing-It-Up Quiz,* further remediation may be necessary.

Mean and Median

Use with Lesson 8-4, exercises 1–8, text pages 264–265.

DIAGNOSTIC INTERVIEW

Before beginning this *Intervention Workshop* lesson, have students complete the following exercises.

Diagnostic Exercises

◆ **Say:** *Write these sets of data and find the mean.*

    7, 5, 6, 3, 4, 8, 2, 6, 5, 4   2, 9, 9, 6, 5, 7, 3, 8, 2, 9   (5; 6)

◆ **Say:** *Write these sets of data and find the median.*

    8, 6, 7, 5, 9, 10, 3, 0, 5   12, 13, 14, 12, 24, 23, 11   (6; 13)

If students answer the Diagnostic Exercises correctly, there is probably no need for them to do this *Intervention Workshop* lesson. Have those students rejoin their class on page 265 of the student textbook. For students who have had difficulty answering the exercises above, continue the *Intervention Workshop*.

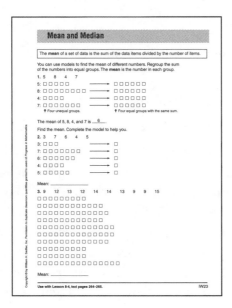

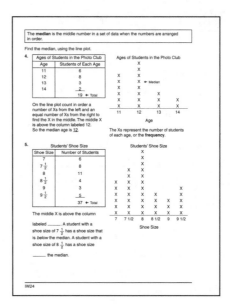

BACKGROUND

This *Intervention Workshop* lesson addresses the problem of distinguishing between the *mean* and the *median*. Students begin by using pictures of tiles to model the data and form equal groups. The number of tiles in each equal group is the *mean*. Then students use a line plot to find the median of a set of data. By forming equal groups and interpreting a line plot students are provided with clear visual models to distinguish the mean from the median.

TEACHING THE *INTERVENTION WORKSHOP* LESSON
Getting Started

Materials: counters

◆ **Say:** *The* mean *of a set of data is the sum of the numbers divided by the number of numbers. For the data set 5, 7, and 3, the sum is 5 + 7 + 3 = 15; the sum divided by the number of numbers is 15 ÷ 3 = 5. The mean of 5, 7, and 3 is 5.*

◆ **Say:** *How can you create a data set of 5 items with a mean of 5?* (Find any five numbers with a sum of 25, since 25 ÷ 5 = 5; sample sets: 2, 3, 3, 7, and 10; 5, 5, 5, 5, and 5.)

◆ **Say:** *The* median *of a set of data is the number in the middle when the numbers in the data set are arranged in order, from least to greatest. For the data set 6, 27, and 3, the data items in order are 3, 6, and 27. The median of the data set 6, 27, and 3 is 6.*

◆ **Say:** *Create a data set of 5 items with a median of 12.* (sample set: 10, 11, 12, 13, and 14)

⑦ USING PAGES IW23 and IW24

Have a student read aloud the material in the box at the top of page IW23. Relate the definition given to the first data set students created earlier in *Getting Started*.

Exercises 1–3 on page IW23
Have a student read aloud the direction line above exercise 1. Guide students through the explanation the following way.

◆ **Say:** *You make 4 equal groups because the number of groups must remain 4. After you make 4 equal groups, the number of tiles in each equal group is the mean.*

◆ **Say:** *You may want to show the data items 5, 8, 4, and 7 with counters, so you can move counters from one group to another until you have 4 equal groups of 6 counters.*

Then have a student read aloud the direction line above exercises 2–3.

You may want to suggest that students use counters.

◆ **Say:** *In these two exercises you draw your own equal groups to find the mean.*

By modeling with tiles exercises 1–3 lead students to analyze the meaning of *mean*. Students gain the understanding that the *mean* stands for a regrouping or balancing of the quantity of items in each set so that the number of items in each set becomes equal, while the total remains the same. Thus the mean can be used as one number to stand for the whole set of numbers.

Exercises 4–5 on page IW24
Have a student read aloud the material in the box at the top of page IW24. Relate that definition to the second data set given as a possible student response in *Getting Started* in the following way.

◆ Say: *Tell how you could make that data set for the median.* (Write the median 12 and then write two numbers less than 12 and two numbers greater than 12.)

Then have a student read aloud the direction line above exercises 4–5.

◆ Say: *The data in each table are displayed in the line plot to the right. A line plot shows data items in order, from least to greatest, so there is no need for you to write the data items in order as a separate step.*

Guide students as necessary to complete exercises 4–5. If students have difficulty counting the Xs to find the median, have them begin on the left at the top of the first column of Xs, count down, then move to the next column, and count up. They should continue counting in this manner. They can start on the right and count in the same way to find the middle X to read the scale for the median.

Have students complete exercise 4.

◆ Say: *The meaning of the median of this set of data can be described in this way: Students aged 11 are below the median age; students aged 13 or 14 are above the median age.*

Have students complete exercise 5.

SUMMING-IT-UP QUIZ

Give this quiz to students.

◆ Say: *Write this set of data and find the mean.*

   60, 75, 70, 95, 70, 50   (70)

◆ Say: *Write the following set of data. Find and then compare the mean and the median.*

   14, 12, 14, 8, 32
(mean: 16; median: 14; The median is less than the mean.)

If students answer the *Summing-It-Up Quiz* correctly, then they should return to Lesson 8-4 and resume work on page 265 of their textbook. If students do not successfully complete the *Summing-It-Up Quiz*, further remediation may be necessary.

Answers: Pages IW23 and IW24

2. 5   **3.** 12   **5.** 8; above

Choose the Appropriate Scale

Use with Lesson 8-7, exercises 10–13, text pages 270–271.

DIAGNOSTIC INTERVIEW

Before beginning this *Intervention Workshop,* have students complete the following exercises.

Diagnostic Exercises

Display this graph on a transparency on the overhead. Explain that the graph shows the runs-batted-in totals for baseball player Mark McGwire for the years 1995 through 1999.

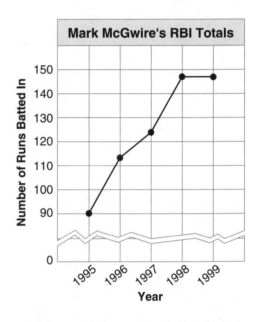

◆ **Say:** *What vertical scale was used in this graph? In other words, what does each interval from line to line equal?* (10) *What does the break in the scale stand for?* (multiples of 10 from 10 to 80) *What reason could there be for using a broken scale in this graph?* (Possible responses: Because there are no data less than 90; to save space.) *What would happen to the graph if you included the intervals that are multiples of 10 from 10 to 80?* (The graph would take up more space.)

Display the following table on a transparency on the overhead or draw this table on the board. Explain that it shows the tickets sold to a new play during the play's first six weeks.

Tickets Sold to *Crazy Like Me*						
Week	1	2	3	4	5	6
Tickets	749	710	820	890	999	941

◆ **Say:** *Suppose you wanted to make a graph of these data. Could you use a broken scale for this graph? Why or why not?* (Yes; there are no data between 0 and 710.) *What is the least number of tickets you need to show?* (710) *What is the greatest number of tickets you need to show?* (999) *Would intervals that were multiples of 5 be a good scale to show these data, which are numbers between 710 and 999? Why or why not?* (No; the graph would be too large.)

◆ **Say:** *Would intervals that were multiples of 500 be a good scale for these data? Why or why not?* (Possible response: No; it would be difficult to plot the points, and the graph would make it seem that there was little difference among the items of data.)

◆ **Say:** *What would be a practical scale for this graph? Explain your answer.* (possible response: a scale with intervals that are multiples of 50, with a break that omits multiples of 50 from 50 to 650; the graph would only take seven intervals, 700, 750, 800, 850, 900, 950, 1000, but it would still be easy to plot the points and to see the differences among the data)

If students answer the Diagnostic Exercises correctly, there is probably no need for them to do this *Intervention Workshop* lesson. Have those students rejoin their class on page 271. If, however, students have difficulty with the exercises above, you should first assign the review lesson on page 17 of *Skills Update* in the student textbook before proceeding with this *Intervention Workshop* lesson.

BACKGROUND

Exercises 10-13 on page 271 require students to make line graphs for sets of data. These *Intervention Workshop* activities address the needs of students who have not yet mastered the skill of choosing an appropriate scale for a line graph. In Activity One students choose an appropriate scale for a line graph. In Activity Two students create a broken scale for a line graph. Activity Three provides further practice in choosing a scale for a line graph, and in deciding whether or not to use a broken scale. Use one or more of these exercises with your students, as needed.

TEACHING THE *INTERVENTION WORKSHOP* LESSON

Getting Started

◆ **Say:** *What are multiples of 5, from 100 to 130?* (100, 105, 110, 115, 120, 125, 130) *What are multiples of 20, from 240 to 340?* (240, 260, 280, 300, 320, 340) *What are multiples of 50, from 1000 to 1200?* (1000, 1050, 1100, 1150, 1200)

ACTIVITY ONE: Choose an Appropriate Scale for a Line Graph

Materials: centimeter grid paper (BLM 1), straightedge for each student

Display the following table on a transparency on the overhead or draw this table on the board. Explain that it shows the number of miles run by a person who is training for a race.

Miles Run by Mario						
Week	1	2	3	4	5	6
Miles Run	15	13	20	22	32	5

Distribute grid paper and straightedges. Draw the axes and labels shown at the right on a transparency on the overhead,

as students do the same on centimeter graph paper at their desks. Have students draw the grid so that the horizontal axis is near the bottom the paper.

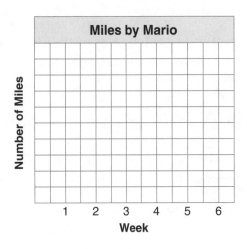

◆ **Say:** *Suppose you want to make a graph of the miles run by Mario. What is the greatest number of miles you need to show?* (32) *How many intervals would you need for a scale that showed intervals of 1, if there was no break in the scale?* (32, the numbers 1 through 32) *Could you show the data clearly on a graph with intervals of 1?* (Yes) *Explain.* (Possible response: Intervals of 1 make it easy to plot and read the data.)

Have students work independently to decide if they could fit a graph with intervals of 1 from 0 to 32 on the vertical axis.

◆ **Say:** *Why are intervals of 1 impractical for this graph?* (With intervals of 1, the graph is too large for the graph paper.)

Now have students consider using intervals of 5.

◆ **Say:** *Suppose you chose a scale with intervals that were multiples of 5. In order to show all the data, what numbers would you need to show above zero on the vertical axis?* (5, 10, 15, 20, 25, 30, 35)

◆ **Say:** *If you used intervals of 5, could you fit the graph on your piece of graph paper?* (Yes) *Could you show the data clearly on a graph with intervals of 5?* (Possible response: Yes, because each piece of data is a multiple of 5, close to a multiple of 5, or midway between two multiples of 5.)

◆ **Say:** *Would someone reading the graph still be able to see the differences among* the items of data? (Yes) *Would intervals of 5 be practical or impractical to use? Explain.* (Possible response: Intervals of 5 would be practical. The graph would fit on the graph paper, it would be easy to plot the points and see the differences among the data.)

Now have students consider using intervals of 10.

◆ **Say:** *Suppose you chose a scale with intervals that were multiples of 10. In order to show all the data, what numbers would you show above zero on the vertical axis?* (10, 20, 30, 40)

◆ **Say:** *If you used intervals of 10, could you fit the graph on your piece of graph paper?* (Yes) *Could you show the data clearly on a graph with intervals of 10?* (Possible response: Not as clearly. It would be harder to see the difference between numbers such as 15 and 13.)

◆ **Say:** *Of these three scales—intervals of 1, 5, or 10—which is the most practical for you to use?* (intervals of 5)

Have students work independently to decide if there is another scale that would be practical to use for the data given.

◆ **Say:** *Can you think of another scale that would be practical to use? Explain.* (Possible response: Intervals of 2, because the graph would fit on the piece of paper and clearly show the differences among the data.)

ACTIVITY TWO: Use a Broken Scale for a Line Graph

Materials: centimeter grid paper (BLM 1), straightedge for each student

Display the following table on a transparency on the overhead. Explain that it shows the number of points scored by basketball star Michael Jordan in each of six seasons.

Michael Jordan's Point Totals						
Season	1986–1987	1987–1988	1988–1989	1989–1990	1990–1991	1991–1992
Points Scored	3041	2868	2633	2753	2580	2404

Distribute grid paper and straightedges. Draw the axes and labels shown below on the transparency on the overhead, as students do the same on centimeter graph paper at their desks. Have students draw the grid so that the horizontal axis is near the bottom the paper.

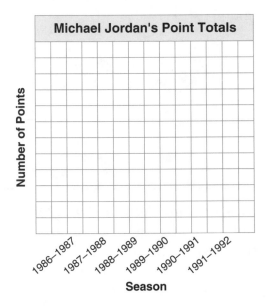

◆ **Say:** *Suppose you want to make a graph of the data shown in the table. What is the least number of points you need to show?* (2404) *What is the greatest number of points you need to show?* (3041) *So you need a scale that allows you to plot numbers from 2400 to 3100.*

Have students think about scales they could use that would allow them to plot numbers between 2400 and 3100 but would still show the differences among the data and not take up too much space.

◆ **Say:** *Would intervals of 1 be practical for plotting data that fall between 2400*
to 3100? Explain. (No. Possible response: With intervals of 1, the intervals from 2400 to 3100 would not fit on the graph paper.)

◆ **Say:** *Would intervals that were multiples of 10 be practical for plotting data that fall between 2400 to 3100? Explain.* (No. Possible response: With intervals of 10, the intervals from 2400 to 3100 would not fit on the graph paper.)

◆ **Say:** *Would intervals that were multiples of 100 be practical for plotting data that fall between 2400 to 3100? Explain.* (Yes. Possible response: With intervals of 100, the intervals from 2400 to 3100 would fit on the graph paper, it would be easy to plot the points, and someone reading the graph would be able to see the approximate differences among the data.)

◆ **Say:** *Suppose you used intervals of 500. To show numbers between 2400 and 3100, you would need the intervals 2000, 2500, 3000, and 3500. Would intervals that were multiples of 500 be practical for plotting data that lie between 2000 and 3500? Explain.* (Possible response: No; although intervals of 500 from 2000 to 3500 would fit on the graph paper, it would be difficult to plot the points, and the graph would make it seem that there was little difference among the data.)

◆ **Say:** *Of these four scales—intervals of 1, 10, 100, or 500—which is the most practical for you to use?* (intervals of 100)

Remind students that although the data range from 2404 to 3104, the number axis still must start at 0.

◆ **Say:** *Suppose you use intervals of 100. Could you fit all of the intervals from zero to 3100 on your piece of graph paper?* (No)

Explain that to save space and fit the graph on the paper, students can use a *broken scale*. The break may come between zero and the least interval in which there is a piece of data.

◆ **Say:** *What is the least number of points that you have to show for any season?* (2404) *If you use intervals of 100, between what two numbers does 2404 fall?* (2400 and 2500) *So since there are no data less than 2404, you can break the scale and leave out intervals that show numbers from zero to 2400.*

Have students draw a break in the scale as shown, then continue by showing intervals of 100 from 2400 to 3100. When students are finished, check their work.

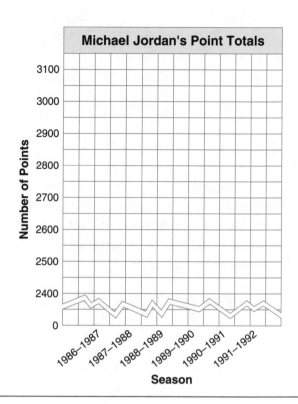

ACTIVITY THREE: Practice Choosing Scales

Materials: centimeter grid paper (BLM 1), straightedge for each student

Display the following table on a transparency on the overhead. Explain that it shows the number of tickets sold to Water World Theme Park each day during one week.

Tickets Sold to Water World Theme Park							
Day	Sun.	Mon.	Tues.	Wed.	Thurs.	Fri.	Sat.
Tickets Sold	505	382	425	404	451	542	598

Have students work with partners to choose a scale and decide if they need a break in a scale. You may wish to suggest different scales for students to consider, such as intervals of 10, 50, 100, and 200. When students are finished, discuss their results

◆ **Say:** *What scale did you choose? Why?* (Possible response: Intervals of 50, because the graph would fit on the piece of paper, be easy to plot data on, and would still show the differences among the data.)

◆ **Say:** *Did you use a break in the scale? Why?* (Possible responses: Yes, because there were no data between 0 and 350, and the graph had to fit on the sheet of paper; No, because the graph fit on the page without putting a break in the scale.)

Have students who used different scales discuss the differences between their graphs. You may wish to have students redraw the graph using a different scale from the first one they chose.

SUMMING-IT-UP QUIZ

Give this quiz to students.

Draw this table on the board. Explain that it shows the pounds of newspaper recycled by a school during the first six months of the school year.

Newspaper Recycled at Tuscan School						
Month	Sept.	Oct.	Nov.	Dec.	Jan.	Feb.
Pounds of Newspaper	135	151	142	108	161	174

◆ **Say:** *Suppose you wanted to make a graph of these data. Could you use a broken scale for this graph? Why or why not?* (Yes; there are no data between 0 and 108.)

◆ **Say:** *Would intervals that were multiples of 1 be a practical scale for this graph? Why or why not?* (No; the graph would be too large.)

◆ **Say:** *Would intervals that were multiples of 100 be a practical scale for this graph? Why or why not?* (No; it would be difficult to plot the points, and the graph would make it seem that there was little difference among the data.)

◆ **Say:** *What would be a practical scale for this graph? Explain your answer.* (possible response: a scale of multiples of 10, with a break that omits multiples of 10 from 10 through 90; the graph would only take nine intervals—100, 110, 120, 130, 140, 150, 160, 170, 180—but it would still be easy to plot the points and to see the differences among the data)

If students answer the *Summing-It-Up Quiz* correctly, They should return to Lesson 8-7 and resume work on page 271 of their textbook. If students do not successfully complete the *Summing-It-Up Quiz,* further remediation may be necessary.

Probability As A Percent Between 0% and 100%

Extension lesson for the gifted and talented

Use with Lesson 8-11, exercises 23–26, text pages 278–279.

BACKGROUND

Exercises 23–26 on page 279 require students to find probabilities for events, including events that are certain or impossible. These *Intervention Workshop* activities provide an opportunity to extend students' skills expressing probability by introducing the concept of expressing probability as a percent. In Activity One students use percents to describe the probabilities of certain and impossible events. In Activity Two students use equivalent fractions and the meaning of percent to rewrite fractional probabilities as percents. In Activity Three students rename fractional probabilities as decimals, then write each decimal as a percent. Use one or more of these activities with your students.

TEACHING THE *INTERVENTION WORKSHOP* LESSON

Getting Started

Draw this spinner on the board.

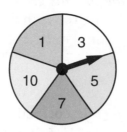

◆ **Say:** *How many possible outcomes are there when you spin this spinner?* (5)

◆ **Say:** *Is spinning a 3 a possible outcome?* (Yes) *What is the probability of spinning a 3?* (1/5)

◆ **Say:** *Is spinning an 8 a possible outcome?* (No) *What is the probability of spinning an 8?* (0/5, or 0)

◆ **Say:** *Is spinning an odd number a possible outcome?* (Yes) *What is the probability of spinning an odd number?* (4/5)

◆ **Say:** *What is the probability of spinning a number between 1 and 10?* (3/5)

Have students create and answer additional probability questions for the spinner.

ACTIVITY ONE: Use Percent to Describe the Probability of Certain and Impossible Events

Draw number cards for 4, 2, 8, 6, and 9 on the board.

◆ **Say:** *What is the probability of picking a number less than 10 from these five cards?* (5/5, or 1) *Suppose you randomly picked a card from this set 100 times, replacing the card after each pick. How many times would you pick a number less than 10?* (100) *Remember: Percent means part of each hundred. In percent, what is the probability of picking a number less than 10?* (100%)

Remind students that an event with a probability of 1, or 100%, is certain to occur. Have students give examples of other events that are certain if they use this set of number cards.

Next, discuss the concept of impossible events.

◆ **Say:** *What is the probability of picking a 1 from these five cards?* (0/5, or 0) *Suppose you randomly picked a card from this set 100 times, replacing the card after each pick. How many times would you pick a 1?* (0) *Remember: Percent means part of each hundred. In percent, what is the probability of picking a 1?* (0%)

ACTIVITY TWO: Use Equivalent Fractions to Rename Probability as Percent

Draw this spinner on the board.

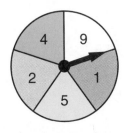

◆ **Say:** *How many possible outcomes are there when you spin this spinner?* (5) *How many odd numbers are on the spinner?* (3) *When you spin this spinner, what fraction describes the probability of spinning an odd number?* (3/5)

Tell students that they will now learn a method for expressing probability as a percent. Remind students that percent means part of 100.

◆ **Say:** *A probability of 3/5 means that you can expect an event to occur in the* ratio of 3 out of every 5 times. A percent is a part of 100. So to express 3/5 as a percent, you can rewrite 3/5 as an equivalent fraction with a denominator of 100. How can you rewrite 3/5 as an equivalent fraction with a denominator of 100?* (Multiply the numerator and denominator of 3/5 by 20.)

Work with students if necessary to find the equivalent fraction.

◆ **Say:** *What equivalent fraction did you find?* (60/100) *What percent has the same meaning as 60 hundredths?* (60%) *What is the percent probability of an event if the probability as a fraction is 3/5?* (60%)

Have students use the spinner pictured on the board to find the probability in percent of spinning the following: an even number; a number less than 9; a 7. (40%; 80%; 0%)

ACTIVITY THREE: Use Decimals to Rename Probability as Percent

Review the connection between decimals and percents.

◆ **Say:** *How can you rewrite a decimal, such as 0.75, as a percent?* (Move the decimal point two places to the right and write a percent symbol.)

◆ **Say:** *Suppose you have a probability that is expressed as a fraction, such as 3/4. How can you use decimals to help you write that probability as a percent?* (Find a decimal that is equivalent to that fraction. Move the decimal point two places to the right and write the percent symbol.)

Draw this spinner on the board.

◆ **Say:** *How many possible outcomes are there when you spin this spinner?* (8) *How many even numbers are on the spinner?* (3) *When you spin this spinner, what fraction describes the probability of spinning an even number?* (3/8)

Now have students write a decimal for 3/8. Remind students that to rename a fraction as a decimal, they can divide the numerator by the denominator.

◆ **Say:** *What decimal can you write for 3/8?* (0.375) *What percent has the same meaning as 3/8?* (37.5%) *What is the percent probability of an event if the probability as a fraction is 3/8?* (37.5%)

Have students use the spinner pictured on the board to find the probability in percent of spinning the following: an odd number; a number less than 5; a 6. (62.5%; 37.5%; 12.5%)

Dependent and Independent Events

Use with Lesson 8-12, exercises 15–18, text pages 280–281.

DIAGNOSTIC INTERVIEW

Before beginning this *Intervention Workshop* lesson, have students complete the following exercises.

Diagnostic Exercises

Draw these letter tiles on the board

◆ **Say:** *Suppose you pick one of these letter tiles from a bag without looking. Then you return the letter tile to the bag and pick again. Does the outcome of your first pick affect the outcome of your second pick?* (No) *What is the probability of picking a* C *on your first pick?* (2/5) *If you return your first pick to the bag, what is the probability of picking a* D *on your second pick?* (1/5) *If you pick a tile, return it to the bag, then pick again, what is the probability of picking a* C*, then a* D? (2/25) *How did you find that probability?* (Possible responses: Found the probability of each event and multiplied the probabilities; used tree diagrams.)

Use the same group of letter tiles to complete the following exercise.

◆ **Say:** *Suppose you pick one of these letter tiles from a bag without looking. You do not return the tile to the bag. Then you pick another tile. Does the outcome of your first pick affect the outcome of your second pick?* (Yes) *What is the probability of picking an* A *on your first pick?* (2/5)

◆ **Say:** *If you pick an* A *tile from the bag on your first pick and do not return the* A *tile to the bag, what is the probability of picking another* A *on your second pick?* (1/4)

◆ **Say:** *If you pick a tile, keep the tile, then pick again, what is the probability of picking an* A*, then another* A? (2/20, or 1/10) *How did you find that probability?* (Possible responses: Found the probability of each event and multiplied the probabilities; used tree diagrams.)

If students answer the Diagnostic Exercises correctly, there is probably no need for them to do this *Intervention Workshop* lesson. Have those students rejoin their class on page 281 of the student textbook. For students who have had difficulties answering the exercises above, continue the *Intervention Workshop*.

BACKGROUND

Exercises 15–18 on page 281 of the student textbook require students to find the probability of independent and dependent compound events. Students who have difficulty with these exercises may need to work on distinguishing between independent and dependent events, finding the probability of each event in a compound event, and using the individual probability of each event in a compound event to find the probability of the compound event. In Activity One students distinguish between independent and dependent events. In Activity Two students use a sample space to find the probability of independent events. In Activity Three students use a sample space to find the probability of dependent events. Use one or more of these activities with your students, as needed.

TEACHING THE *INTERVENTION WORKSHOP* LESSON

Getting Started

Draw this diagram on the board. Explain that each square represents a cube of the color indicated.

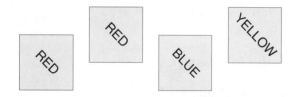

◆ **Say:** *Suppose you pick a cube without looking. How many possible outcomes are there?* (4) *How many red cubes are there?* (2) *What is the probability of picking a red cube?* (2/4, or 1/2)

ACTIVITY ONE: Distinguish Between Independent and Dependent Events

Materials: red, blue, and green connecting cubes; paper bag

Have students place 2 red connecting cubes, 1 blue connecting cube, and 2 green connecting cubes in a bag.

◆ **Say:** *How many cubes are in the bag?* (5) *How many red cubes are in the bag?* (2) *Suppose you pick a cube without*

looking. What is the probability of picking a red cube? (2/5)

◆ **Say:** *Suppose you pick a red cube from the bag. If you return the cube to the bag and pick again, what is the probability of picking a red cube again?* (2/5) *Did the outcome of the first pick affect the probability of the second event?* (No)

Why did the first pick not change the probability of picking red in the second pick? (Because the first pick was returned to the bag, the same cubes were left to choose from in the second pick.)

Remind students that when the outcome of one event does not affect the second event, the events are called *independent events*.

Next discuss the concept of *dependent events*.

◆ **Say:** *The probability of picking a red cube is 2/5. Suppose you pick a red cube and do not return the cube to the bag. For your second pick, how many cubes are left in the bag?* (4) *How many red cubes are left in the bag?* (1) *For your second pick, what is the probability of picking red?* (1/4)

◆ **Say:** *Did the outcome of the first pick affect the probability of the second event?* (Yes) *Why did the first pick change the probability of picking red in the second pick?* (The first pick decreased the number of cubes left to choose from in the second pick.)

Remind students that when the outcome of one event affects the second event, the events are called *dependent events*.

ACTIVITY TWO: Use a Sample Space to Find the Probability of Independent Events

Draw these letter tiles on the board.

F	E	F	A	C

D	A	M	I	D

◆ **Say:** *Suppose you are going to pick a letter tile without looking, put it back, then pick another letter tile. You want to know the probability of picking an* A, *then an* M. *You can use a sample space to help you find the possible picks and the probability for each pick. Then you can use those probabilities to find the probability of picking an* A, *then an* M.

Explain that a sample space is a list of all of the possible outcomes for a single event. Have students write a sample space for the first pick as you do so on the board. Then lead students to use the sample space to calculate the probability of picking an *A* on the first pick.

◆ **Say:** *You want to pick an* A *on your first pick. How many* A*'s are there?* (2) *How many letter tiles are there in all?* (10) *What is the probability of picking an* A *on your first pick?* (2/10, or 1/5)

> Sample Space for First Pick:
>
> F E F A C
>
> D A M I D
>
> Favorable outcome(s): A, A
>
> Probability: $P(A) = \dfrac{2}{10} = \dfrac{1}{5}$

Now help students calculate the probability of picking an *M* on the second pick.

◆ **Say:** *Suppose you pick an* A *on your first pick. Then you put the tile back. Make a sample space that shows the letters available for your second pick.*

As students make a sample space for the second pick, do the same on the board. Then lead students to use the sample space to calculate the probability of picking an *M* on the second pick.

Sample Space for Second Pick:

F   E   F   A   C

D   A   M   I   D

Favorable outcome(s):   M

Probability: $P(M) = \dfrac{1}{10}$

◆ **Say:** *You want to pick an* M *on your second pick. How many* Ms *are there?* (1) *How many letter tiles are left to pick from?* (10) *What is the probability of picking an* M *on your second pick?* (1/10)

As students answer your questions, write the answers on the board as shown.

Now have students find the probability of picking an *A*, then an *M*, if they return the *A* to the letter tiles after picking it. Remind students that to find the probability that one event, then another, will occur, they can multiply the probabilities of each event. Write the following on the board:

$$P(A, M) = 1/5 \times 1/10$$

◆ **Say:** *Suppose you pick a letter tile without looking, put it back, and pick again. What is the probability of picking an* A, *then an* M? (1/50)

ACTIVITY THREE: Use a Sample Space to Find the Probability of Dependent Events

Draw these letter tiles on the board.

A   E   G   B   I

A   D   F   E   K

L   A   C   I   O

◆ **Say:** *Suppose you are going to pick a letter tile without looking, then pick another letter tile. You are going to keep both letter tiles. You want to know the probability of picking an* A, *then a consonant. You can use a sample space to help you find the possible picks and the* probability *for each pick. Then you can use those probabilities to find the probability of picking an* A, *then a consonant.*

Remind students that a sample space is a list of all of the possible outcomes for a single event. Have students write a sample space for the first pick as you do so on the board. Then lead students to use the sample space to calculate the probability of picking an *A* on the first pick.

◆ **Say:** *You want to pick an* A *on your first pick. How many* A's *are there?* (3) *How many letter tiles are there in all?* (15) *What is the probability of picking an* A *on your first pick?* (3/15, or 1/5)

As students answer your questions, write the responses on the board as shown.

Sample Space for First Pick:

A E G B I

A D F E K

L A C I O

Favorable outcome(s): A, A, A

Probability: $P(A) = \dfrac{3}{15} = \dfrac{1}{5}$

Now help students calculate the probability of picking a consonant on the second pick.

◆ **Say:** *Suppose you pick an* A *on your first pick. You keep the* A. *Make a sample space that shows the letters that are left for the second pick.*

As students make a sample space for the second pick, do the same on the board. Then lead students to use the sample space to calculate the probability of picking a consonant on the second pick.

◆ **Say:** *You have picked an* A *on your first pick. You want to pick a consonant on your second pick. How many letter tiles are left for your second pick?* (14) *What consonants are there?* (G, B, D, F, K, L, C) *How many consonants is that?* (7) *What is the probability of picking a consonant on your second pick?* (7/14, or 1/2)

As students answer your questions, write the responses on the board as shown.

Sample Space for Second Pick:

  E G B I

A D F E K

L A C I O

Favorable outcome(s): G, B, D, F, K, L, C

Probability: $P(\text{consonant}) = \dfrac{7}{14} = \dfrac{1}{2}$

Now have students find the probability of picking an *A*, then a consonant. Remind students that to find the probability that one event, then another, will occur, they can multiply the probabilities of each event. Write the following on the board.

$$P(A, \text{consonant}) = 1/5 \times 1/2$$

◆ **Say:** *Suppose you pick a letter tile, keep it, then pick another letter tile. What is the probability of picking an* A, *then a consonant?* (1/10)

SUMMING-IT-UP QUIZ

Give this quiz to students.

Draw these number cards on the board.

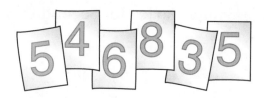

◆ **Say:** *Suppose you pick one of these number cards, replace the card, then pick again. Does the outcome of the first pick affect the outcome of the second pick?* (No) *If you replace your first pick, what is the probability of picking a 5, then a 6?* (1/18)

◆ **Say:** *Suppose you pick one of these number cards, keep it, then pick another number card. Does the outcome of the first pick affect the outcome of the second event?* (Yes) *If you keep the first card, what is the probability of picking a 5, then a 6?* (1/3 × 1/5, or 1/15)

If students answer the *Summing-It-Up Quiz* correctly, they should then resume work on page 281 of their textbook. If students do not successfully complete the *Summing-It-Up Quiz*, further remediation may be necessary.

Introduction to the Protractor

Use with Lesson 9-3, exercises 7–18, text pages 300–301.

DIAGNOSTIC INTERVIEW

Before beginning this *Intervention Workshop,* have students complete the following exercises.

Diagnostic Exercises

Make a transparency with ∠*ABC* of 30° that opens right and ∠*XMY* of 110° that opens left. Display the transparency on the overhead.

◆ **Say:** *What is the vertex of angle ABC?* (B) *To measure angle ABC, what part of the protractor goes on the vertex?* (the center of the protractor) *Where should ray BC point on the protractor?* (to the 0° mark on the right side of the protractor)

Place the protractor on the transparency as shown, so that it measures ∠*ABC.*

◆ **Say:** *What is the measure of angle ABC?* (30°) *Which scale did you use to get that measurement: the outer scale or the inner scale?* (the inner scale)

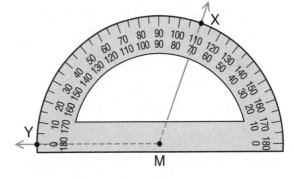

Now point out ∠*XMY* on the overhead.

◆ **Say:** *What is the vertex of angle XMY?* (M) *To measure angle XMY, what part of the protractor goes on the vertex?* (the center of the protractor) *Where should ray MY point on the protractor?* (to the 0° mark on the left side of the protractor)

Place the protractor on the transparency as shown, so that it measures ∠*XMY.*

◆ **Say:** *What is the measure of angle XMY?* (110°) *Which scale did you use to get that measurement: the outer scale or the inner scale?* (the outer scale)

If students answer the Diagnostic Exercises correctly, there is probably no need for them to do this *Intervention Workshop* lesson. Have those students rejoin their class on page 301 of the student textbook. For students who have had difficulties answering the exercises above, continue the *Intervention Workshop.*

BACKGROUND

Exercises 7–18 on page 301 of the student textbook require students to draw angles of given measure and to find the measures of different angles. These *Intervention Workshop* activities address the needs of students who still have difficulty using a protractor. In Activity One students use the intersection of the base ray of the angle and the 0° mark to choose the appropriate protractor scale and measure angles. In Activity Two students compare angles to right angles in order to choose the appropriate protractor scale and measure angles. In Activity Three students use a protractor to draw angles of given measures. Use one or more of these activities with your students, as needed.

TEACHING THE *INTERVENTION WORKSHOP* LESSON

Getting Started

Make a transparency with ∠*FBN* of 50° that opens to the right, ∠*CVP* of 110° that opens to the left, and ∠*RST* of 90° that opens to the right. Put the transparency on the overhead.

◆ **Say:** *Which angle is a right angle?* (∠*RST*) *Which angle is less than a right angle?* (∠*FBN*) *Which angle is greater than a right angle?* (∠*CVP*)

ACTIVITY ONE: Use the Base Ray of the Angle to Choose the Appropriate Protractor Scale

Materials: protractor; angle worksheet I (BLM 10) for each student; transparency of angle worksheet I (BLM 10)

Make a transparency of the worksheet and demonstrate how to use to use the protractor to measure angles. Start with ∠*ERG*, which opens to the right.

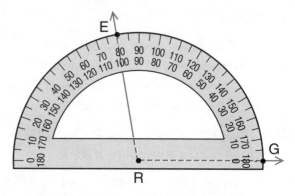

◆ **Say:** *What is the vertex of angle ERG?* (R) *To measure angle ERG, what part of the protractor goes on the vertex?* (the center of the protractor) *Where should ray RG point on the protractor?* (to the 0° mark on the right side of the protractor)

Place the protractor on the transparency as shown, so that it measures ∠*ERG*. Have students do the same on their worksheets.

◆ **Say:** *Look where ray RG meets the zero-degree mark. Is the zero-degree mark on the outer scale or inner scale?* (the inner scale) *Use that same scale to measure angle ERG. What number does ray RE meet on that scale?* (100) *What is the measure of angle ERG?* (100°)

Now demonstrate how to use to use the protractor to measure an angle that opens to the left, ∠*QMT*.

◆ **Say:** *What is the vertex of angle QMT?* (M) *To measure angle QMT, what part of the protractor goes on the vertex?* (the center of the protractor) *Where should ray MT point on the protractor?* (to the 0° mark on the left side of the protractor)

Place the protractor on the transparency as shown, so that it measures ∠QMT. Have students do the same on their worksheets.

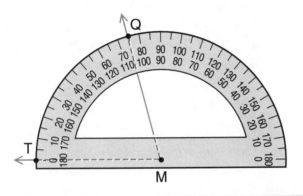

◆ **Say:** *Look where ray MT meets the zero-degree mark. Is the zero-degree mark on the outer scale or inner scale?* (the outer scale) *Use that same scale to measure angle QMT. What number does ray MQ meet on that scale?* (75) *What is the measure of angle QMT?* (75°)

Have students work independently to measure the other four angles on the worksheet. (∠UVW = 45°, ∠LOC = 130°, ∠AXJ = 155°, ∠FWZ = 80°)

ACTIVITY TWO: Use Estimates of Angle Measure to Choose the Appropriate Protractor Scale

Materials: protractor; angle worksheet II (BLM 11) for each student; transparency of angle worksheet II (BLM 11)

Using the transparency of the worksheet, demonstrate how to use estimates of angle measure to choose the appropriate protractor scale and measure. Start with ∠OKN, which is an acute angle.

◆ **Say:** *Is angle OKN greater than a right angle or less than a right angle?* (less than a right angle)

Lead students to see how to place the protractor on the angle in order to get the measurement.

◆ **Say:** *What is the vertex of angle OKN?* (K) *To measure angle OKN, what part of the protractor goes on the vertex?* (the center of the protractor) *Where should ray KN point on the protractor?* (to the 0° mark on the right side of the protractor)

Place the protractor on the transparency as shown, so that it measures ∠OKN. Have students do the same on their worksheets.

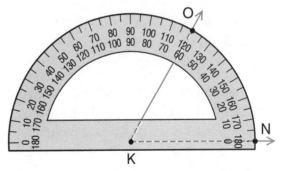

◆ **Say:** *What number does ray KO meet on the inner scale?* (60) *What number does ray KO meet on the outer scale?* (120) *You estimated that angle OKN was less than a right angle. Which measurement gives an angle that is less than a right angle: 60 degrees or 120 degrees?* (60°) *What is the measure of angle OKN?* (60°)

Now demonstrate how to use to use the protractor to measure an obtuse angle.

◆ **Say:** *Is angle* TRL *greater than a right angle or less than a right angle?* (greater than a right angle)

Lead students to see how to place the protractor on the angle in order to get the measurement.

◆ **Say:** *What is the vertex of angle* TRL? (R) *To measure angle* TRL, *what part of the protractor goes on the vertex?* (the center of the protractor) *Where should ray* RL *point to on the protractor?* (the 0° mark on the left side of the protractor)

Place the protractor on the transparency as shown, so that it measures ∠TRL. Have students do the same on their worksheets.

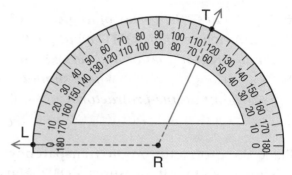

◆ **Say:** *What number does ray* RT *meet on the inner scale?* (65) *What number does ray* RT *meet on the outer scale?* (115) *You estimated that angle* TRL *was greater than a right angle. Which measurement gives an angle that is greater than a right angle: 65 degrees or 115 degrees?* (115°) *What is the measure of angle* TRL? (115°)

Have students work independently to measure the other four angles on the worksheet. (∠CEG = 135°, ∠XVD = 25°, ∠UWA = 40°, ∠SPI = 145°)

ACTIVITY THREE: Use a Protractor to Draw an Angle of a Given Measure

Materials: protractor for each student

Tell students that they will now use a protractor to draw ∠XYZ that is 80° and opens to the right.

On a transparency draw ray *YZ* that points to the right. Have students do the same on a sheet of paper. Then demonstrate how to use the protractor to draw an angle.

◆ **Say:** *Point* Y *is the vertex of the angle you are drawing. What part of the protractor should line up with point* Y? (the center mark) *Since ray* YZ *is the base ray of the angle, to what measurement on the right of the protractor should ray* YZ *point?* (the 0° mark)

Place the protractor on the transparency as shown, so that point *Y* is at the center mark and ray *YZ* passes through the 0° mark. Have students do the same.

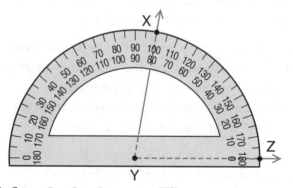

◆ **Say:** *Look where ray* YZ *meets the zero-degree mark. Is the zero-degree mark on the outer scale or inner scale?* (the inner scale) *On that same scale, find the mark for 80 degrees. Draw a point* X *at 80 degrees. Now draw ray* YX. *What is the measure of angle* XYZ? (80°)

Tell students that they will now use a protractor to draw ∠GHJ that is 110° and opens to the left.

On a transparency draw ray *HJ* that points to the left. Have students do the same on a sheet of paper. Then demonstrate how to use the protractor to draw the angle.

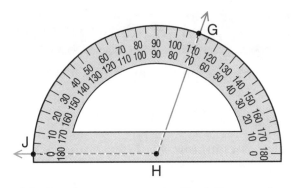

◆ **Say:** *Point* H *is the vertex of the angle you are drawing. What part of the protractor should line up with point* H? (the center mark) *Since ray* HJ *is the base ray of the angle, to what measurement on the left of the protractor should ray* HJ *point?* (the 0° mark)

Place the protractor on the transparency as shown, so that point *H* is at the center mark and ray *HJ* passes through the 0° mark. Have students do the same.

◆ **Say:** *Look where ray* HJ *meets the zero-degree mark. Is the zero-degree mark on the outer scale or inner scale?* (the outer scale) *On that same scale, find the mark for 110 degrees. Draw a point G at 110 degrees. Now draw ray* HG. *What is the measure of angle* GHJ? (110°)

Have students work independently to draw the following angles: ∠*RAD* of 40° that opens to the right, ∠*COB* of 70° that opens to the left, ∠*LCD* of 170° that opens to the right, and ∠*KPJ* of 125° that opens to the left.

SUMMING-IT-UP QUIZ

Give this quiz to students.

Make a transparency with ∠*CBN* of 140° that opens to the right and ∠*QED* of 35° that opens to the left. Display it on the overhead.

Place the protractor on the transparency as shown, so that it measures ∠*CBN*.

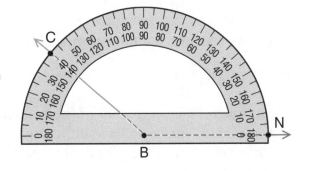

◆ **Say:** *What is the measure of angle* CBN? (140°) *Which scale did you use to get that measurement: the outer scale or inner scale?* (the inner scale)

Now place the protractor on the transparency as shown, so that it measures ∠*QED*.

◆ **Say:** *What is measure of angle* QED? (35°) *Which scale did you use to get that measurement: the outer scale or inner scale?* (the outer scale)

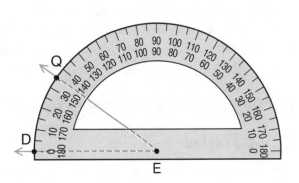

If students answer the *Summing-It-Up Quiz* correctly, they should then resume work on page 301 of their textbook. If students do not successfully complete the *Summing-It-Up Quiz,* further remediation may be necessary.

Draw and Measure Triangles

Use with Lesson 9-7, exercises 9–12, text pages 308–309.

DIAGNOSTIC INTERVIEW

Before beginning this *Intervention Workshop,* have students complete the following exercises.

Diagnostic Exercises

Make a transparency with a right isosceles triangle labeled *A*, an equilateral triangle labeled *B*, and an obtuse scalene triangle labeled *C*. Display the transparency on the overhead. Have students classify the triangles by the lengths of their sides, identifying each triangle as equilateral, isosceles, or scalene.

◆ **Say:** *Which triangle is an equilateral triangle?* (*B*) *Which triangle is a scalene triangle?* (*C*) *Which triangle is an isosceles triangle?* (*A*)

Now have students classify the triangles by the measure of their angles, identifying each triangle as acute, right, or obtuse.

◆ **Say:** *Which triangle is an obtuse triangle?* (*C*) *Which triangle is a right triangle?* (*A*) *Which triangle is an acute triangle?* (*B*)

If students answer the Diagnostic Exercises correctly, there is probably no need for them to do this *Intervention Workshop* lesson. Have those students rejoin their class on page 309 of the student textbook. For students who have had difficulties answering the exercises above, continue the *Intervention Workshop*.

BACKGROUND

Exercises 9–12 on page 309 of the student textbook require students to classify triangles. Students who have difficulty with these exercises may benefit from work in which they draw and measure triangles. In Activity One students use angle measure to draw and classify triangles. In Activity Two students identify congruent sides of triangles and use this information to classify triangles. Activity Three provides students with an opportunity to draw different types of triangles, and to classify triangles according to both side length and angle measure. Use one or more of these activities with your students, as needed.

TEACHING THE *INTERVENTION WORKSHOP* LESSON

Getting Started

Have students draw a few different triangles.

◆ **Say:** *Do any of your triangles have a right angle?* (Answers will vary.) *Do any of your triangles have an angle that is greater than a right angle?* (Answers will vary.) *Do any of your triangles have two sides of the same length?* (Answers will vary.) *Do any have three sides of the same length?* (Answers will vary.)

ACTIVITY ONE: Use Angle Measure to Draw and Classify Triangles

Materials: triangular dot paper (BLM 12), straightedge, protractor for each student

Review angle classification. Make a transparency with an obtuse angle labeled *A*, a right angle labeled *B*, and an acute angle labeled *C*. Display the transparency on the overhead. Have students classify the angles, identifying each angle as acute, right, or obtuse.

◆ **Say:** *Which angle is a right angle?* (*B*) *Which angle is an acute angle?* (*C*) *Which angle is an obtuse angle?* (*A*)

Have students draw segments that meet at a right angle on triangular dot paper as you do so on a transparency of triangular dot paper on the overhead. Then have students draw a segment that joins the endpoints of the first two line segments and forms a triangle as you do the same.

◆ **Say:** *Does this triangle have a right angle?* (Yes) *What do you call a triangle with one right angle: an acute triangle, an obtuse triangle, or a right triangle?* (a right triangle)

◆ **Say:** *Write the label* right triangle *next to this triangle. What are the other two angles in this right triangle: acute angles, obtuse angles, or right angles?* (acute angles)

Have students draw segments that meet at an obtuse angle on triangular dot paper as you do so on a transparency of triangular dot paper on the overhead. Then have students draw a segment that joins the endpoints of the first two line segments and forms a triangle as you do the same.

◆ **Say:** *Does this triangle have an obtuse angle?* (Yes) *What do you call a triangle with one obtuse angle: an acute triangle, an obtuse triangle, or a right triangle?* (an obtuse triangle)

◆ **Say:** *Write the label* obtuse triangle *next to this triangle. What are the other two angles in this obtuse triangle: acute angles, obtuse angles, or right angles?* (acute angles)

Draw an acute triangle on a transparency of triangular dot paper, as shown. Display the transparency on the overhead.

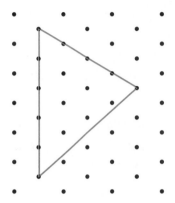

Have students copy the triangle on their triangular dot paper. Point to each angle and ask if the angle is acute, right, or obtuse.

◆ **Say:** *How many acute angles does this triangle have?* (3)

Explain that when all three angles in a triangle are acute, the triangle is called an *acute triangle*. Have students write the label *acute triangle* next to this triangle.

◆ **Say:** *Can a right triangle be an acute triangle? Why or why not?* (No. A right triangle has 1 right angle and 2 acute angles. An acute triangle has 3 acute angles.)

◆ **Say:** *Can a right triangle be an obtuse triangle? Why or why not?* (No. A right triangle does not have any obtuse angles.)

◆ **Say:** *Can an acute triangle be an obtuse triangle? Why or why not?* (No. An acute triangle has only acute angels.)

Have students work independently to draw a right triangle, an acute triangle, and an obtuse triangle that are different from the ones that they have already drawn.

Then have them exchange papers with partners and check each other's work.

ACTIVITY TWO: Use Side Length to Draw and Classify Triangles

Materials: triangular dot paper (BLM 12), centimeter ruler for each student

Draw a scalene triangle on a transparency of triangular dot paper, as shown. Display the transparency on the overhead. Have students copy the triangle on their triangular dot paper. If necessary, review the meaning of the term *congruent*.

◆ **Say:** *Does this triangle have all congruent sides, two congruent sides, or no congruent sides?* (no congruent sides) *How do you know?* (Possible response: Using the grid to measure, one can see that there are no congruent sides.)

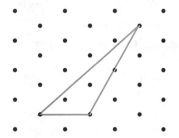

Explain that when a triangle has no congruent sides, the triangle is called a *scalene triangle*. Have students write the label *scalene triangle* next to this triangle.

Draw an isosceles triangle on a transparency of triangular dot paper, as shown. Display the transparency on the overhead. Have students copy the triangle on their triangular dot paper.

◆ **Say:** *Does this triangle have all congruent sides, two congruent sides, or no congruent sides?* (2 congruent sides) *How do you know?* (Possible response: Using the grid to measure, one can see that there are 2 congruent sides.)

Explain that when a triangle has 2 congruent sides, the triangle is called an *isosceles triangle*. Have students write the label *isosceles triangle* next to this triangle.

Draw an equilateral triangle on a transparency of triangular dot paper, as shown. Display the transparency on the overhead. Have students copy the triangle on their triangular dot paper.

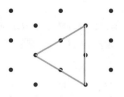

◆ **Say:** *Does this triangle have all congruent sides, two congruent sides, or no congruent sides?* (all congruent sides) *How do you know?* (Possible response: Using the grid to measure, one can see that there are three congruent sides.)

Explain that when a triangle has 3 congruent sides, the triangle is called an *equilateral triangle*. Have students write the label *equilateral triangle* next to this triangle.

Have students work independently to draw a scalene triangle, an isosceles triangle, and an equilateral triangle that are different from the ones that they have already drawn. Then have them exchange papers with partners and check each other's work.

ACTIVITY THREE: Practice Drawing and Classifying Triangles

Materials: triangular dot paper (BLM 12), centimeter ruler, protractor for each student

Draw a right scalene triangle on a transparency of triangular dot paper, as shown. Display the transparency on the overhead. Have students copy the triangle on their triangular dot paper.

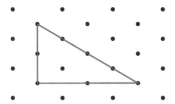

Remind students that there are two ways of classifying any triangle: by the measure of the angles and by the measure of the sides. Then lead them to classify this triangle.

◆ **Say:** *Does this triangle have a right angle, an obtuse angle, or all acute angles?* (a right angle) *Classify this triangle according to the measures of its angles. Is this triangle acute, obtuse, or right?* (right) *Does this triangle have all congruent sides, two congruent sides, or no congruent sides?* (no congruent sides) *Classify this triangle according to the measures of its sides. Is this triangle scalene, isosceles, or equilateral?* (scalene)

Have students work with partners. Each partner draws an equilateral triangle, an isosceles triangle, and a scalene triangle, and labels those triangles according to the measures of their sides. Partners exchange papers and label each other's triangles according to the measures of their angles.

SUMMING-IT-UP QUIZ

Give this quiz to students.

Draw an obtuse isosceles triangle on a transparency of triangular dot paper on the overhead. Have students copy the triangle onto a piece of triangular dot paper.

◆ **Say:** *Does this triangle have a right angle, an obtuse angle, or all acute angles?* (an obtuse angle) *Classify this triangle according to the measures of its angles. Is this triangle acute, obtuse, or right?* (obtuse) *Does this triangle have all congruent sides, two congruent sides, or no congruent sides?* (2 congruent sides) *Classify this triangle according to the measures of its sides. Is this triangle scalene, isosceles, or equilateral?* (isosceles)

If students answer the *Summing-It-Up Quiz* correctly, they should then resume work on page 309 of their textbook. If students do not successfully complete the *Summing-It-Up Quiz*, further remediation may be necessary.

Polygons Other Than Triangles

Use with Lesson 9-8, exercises 1–7, text pages 310–311.

DIAGNOSTIC INTERVIEW

Before beginning this *Intervention Workshop* lesson, have students complete the following exercises.

Diagnostic Exercises

On the board, draw a square labeled *A*, a nonsquare rhombus labeled *B*, and a parallelogram that is not a rhombus labeled *C*.

◆ **Say:** *How are shapes* **A** *and* **B** *alike?* (They both have 4 congruent sides and 2 pairs of parallel sides.) *How are shapes* **A** *and* **B** *different?* (Shape *A* has 4 right angles, and shape *B* has no right angles.) **Which shape is a square?** (Shape *A*). *How are shapes* **B** *and* **C** *different?* (Shape *B* has 4 congruent sides, and shape *C* only has 2 pairs of congruent sides.)

◆ **Say:** *Which shape is a parallelogram, but not a rhombus: shape* B *or* C*?* (*C*) *Which shape is a rhombus: shape* B *or* C*?* (*B*)

Have students draw a shape with 4 sides in which only 1 pair of sides is parallel.

◆ **Say:** *Is your shape a parallelogram, a rectangle, a square, a trapezoid, or a rhombus?* (a trapezoid)

If students answer the Diagnostic Exercises correctly, there is probably no need for them to do this *Intervention Workshop* lesson. Have those students rejoin their class on page 311 of the student textbook. For students who have had difficulties answering the exercises above, continue the *Intervention Workshop*.

BACKGROUND

Exercises 1–7 on page 311 of the student textbook require students to classify quadrilaterals. These *Intervention Workshop* activities address the needs of students who still have difficulty identifying attributes of quadrilaterals and using these attributes to classify quadrilaterals. Activity One provides students with opportunities to identify parallel sides, congruent sides, and right angles and to use these attributes to classify the quadrilaterals. In Activity Two students sort and classify pattern blocks according to their attributes, classifying figures as they sort. Activity Three provides students with an opportunity to draw and classify different types of quadrilaterals. Use one or more of these activities with your students, as needed.

TEACHING THE *INTERVENTION WORKSHOP* LESSON

Getting Started

On the board, draw this figure.

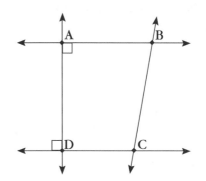

◆ **Say:** *Which lines are parallel?* ($\overleftrightarrow{AB}$ and $\overleftrightarrow{DC}$) *Which lines meet to form right angles?* ($\overleftrightarrow{AB}$ and $\overleftrightarrow{AD}$; $\overleftrightarrow{AD}$ and $\overleftrightarrow{DC}$)

Review the term *congruent*.

◆ **Say:** *What does it mean when you say that two line segments are congruent?* (The two line segments have the same length.)

ACTIVITY ONE: Use Drawings to Identify Attributes of Quadrilaterals

Materials: dot paper (BLM 13), triangular dot paper (BLM 12), blue and red crayons, centimeter ruler for each student

Draw a trapezoid on a transparency of dot paper, as shown. Display the transparency on the overhead. Have students copy the shape on their dot paper.

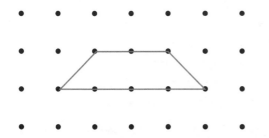

◆ **Say:** *Can you find a pair of parallel sides?* (Yes) *Color each of the parallel sides red. Is the other pair of sides parallel?* (No)

Explain that when a shape has four sides and just one pair of sides is parallel, that shape is called a *trapezoid*. Have students write the label *trapezoid* next to this shape.

Draw a parallelogram on a dot paper transparency, as shown. Display the transparency on the overhead. Have students copy the shape on their dot paper.

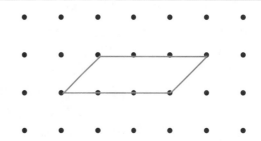

◆ **Say:** *Can you find a pair of parallel sides?* (Yes) *Color this pair of parallel sides red. Is the other pair of sides parallel?* (Yes) *Color that pair of sides blue.*

◆ **Say:** *Are the red sides congruent?* (Yes) *How do you know?* (Possible response: Using the grid to measure, one can see that the sides have the same length.) *Mark each of the red sides with a tick mark to show they are congruent. Are the blue sides congruent?* (Yes) *Mark each of the blue sides with a double tick mark to show they are congruent. Are the red sides congruent to the blue sides?* (No)

Explain that when a shape is made of two pairs of parallel congruent sides, that shape is called a *parallelogram*. Have students write the label *parallelogram* next to this shape.

Draw a rectangle on a dot paper transparency, as shown. Display the transparency on the overhead. Have students copy the shape on their dot paper.

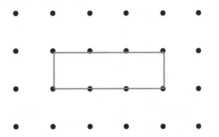

◆ **Say:** *Can you find a pair of parallel sides?* (Yes) *Color each of these two sides red. Is the other pair of sides parallel?* (Yes) *Color that pair of sides blue.*

◆ **Say:** *Are the red sides congruent?* (Yes) *How do you know?* (Possible response: Using the grid to measure, one can see that the sides are the same length.) *Mark each of the red sides with a tick mark to show they are congruent. Are the blue sides congruent?* (Yes) *Mark each of the blue sides with a double tick mark to show they are congruent.*

◆ **Say:** *Can you find any right angles?* (Yes) *Mark each right angle. How many right angles did you find?* (4)

Explain that when a shape is made of two pairs of parallel congruent sides, and has 4 right angles, that shape is called a *rectangle*. Have students write the label *rectangle* next to this shape.

Draw a square on a dot paper transparency, as shown. Display the transparency on the overhead. Have students copy the shape on their dot paper.

◆ **Say:** *Can you find a pair of parallel sides?* (Yes) *Color each of these two sides red. Is the other pair of sides parallel?* (Yes) *Color that pair of sides blue.*

◆ **Say:** *Are the red sides congruent?* (Yes) *How do you know?* (Possible response: Using the grid to measure, one can see that the sides are the same length.) *Are the blue sides congruent?* (Yes) *Are the blue sides congruent to the red sides?* (Yes) *Mark each side with a tick mark to show that all four sides are congruent. Can you find any right angles?* (Yes) *Mark each right angle. How many right angles did you find?* (4)

Explain that when a shape is made of 4 congruent sides, has 2 pairs of parallel sides, and has 4 right angles, that shape is called a *square*. Have students write the label *square* next to this shape.

Draw a rhombus on a triangular dot paper transparency, as shown. Display the transparency on the overhead. Have students get a piece of triangular dot paper, and copy the shape onto this paper.

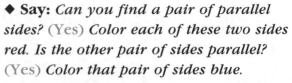

◆ **Say:** *Can you find a pair of parallel sides?* (Yes) *Color each of these two sides red. Is the other pair of sides parallel?* (Yes) *Color that pair of sides blue.*

◆ **Say:** *Are the red sides congruent?* (Yes) *How do you know?* (Possible response: Using the grid to measure, one can see that the sides are the same length.) *Are the blue sides congruent?* (Yes) *Are the blue sides congruent to the red sides?* (Yes) *Mark each side with a tick mark to show that all four sides are congruent. Can you find any right angles?* (No)

Explain that when a shape is made of 4 congruent sides and has 2 pairs of parallel sides, that shape is called a *rhombus*. Have students write the label *rhombus* next to this shape.

Have students work independently to draw a trapezoid, a rhombus that is not a square, a rectangle that is not a square, and a square, all of which are different from the ones they have already drawn. Then have them exchange papers with partners and label each other's shapes.

ACTIVITY TWO: Sort and Classify Pattern Blocks

Materials: pattern blocks, attribute worksheet (BLM 14) for each pair of students, transparency of attribute worksheet

Have students work in pairs. Give the following to each pair of students: an orange square, 2 orange squares taped together to form a rectangle, a blue rhombus, 2 blue rhombuses taped together to form a parallelogram, a red trapezoid.

On the overhead, display a transparency of the attribute worksheet.

Tell students that they will sort their shapes into the columns on the worksheet. Have a volunteer read aloud the description given in the first column of the worksheet as other students read silently.

Trapezoid	Parallelogram	Rectangle	Rhombus	Square
Exactly one pair of parallel sides	Two pairs of sides that are parallel and congruent	Two pairs of parallel congruent sides 4 right angles	Two pairs of parallel sides 4 congruent sides	Two pairs of parallel sides 4 congruent sides 4 right angles

◆ **Say:** *Put any shapes that fit this description in the first column. Which shape or shapes fit this description?* (the red shape) *What is it about the large orange shape that does not fit that description?* (It has two pairs of parallel sides.)

Have a volunteer read aloud the description given in the second column of the worksheet as other students read silently.

◆ **Say:** *Put any shapes that fit this description in the second column. Which shape or shapes fit this description?* (the big blue shape, the small blue shape, the big orange shape, the small orange shape) *What is it about the red shape that does not fit this description?* (It has only 1 pair of parallel sides.)

Have a volunteer read aloud the description given in the third column of the worksheet as other students read silently.

◆ **Say:** *Put any shapes that fit this description in the third column. Take shapes from other columns if you need to. Which shape or shapes fit this description?* (the big orange shape, the small orange shape) *What is it about the blue shapes that does not fit this description?* (They have no right angles.)

Have a volunteer read aloud the description given in the fourth column of the worksheet as other students read silently.

◆ **Say:** *Put any shapes that fit this description in the fourth column. Take shapes from other columns if you need to. Which shape or shapes fit this description?* (the small orange shape, the small blue shape) *What is it about the large blue shape that does not fit this description?* (It has only 2 congruent sides.) *What is it about the large orange shape that does not fit this description?* (It has only 2 congruent sides.)

Have a volunteer read aloud the description given in the fifth column of the worksheet as other students read silently.

◆ **Say:** *Put any shapes that fit this description in the fifth column. Take shapes from other columns if you need to. Which shape or shapes fit this description?* (the small orange shape) *What is it about the blue shapes that do not fit this description?* (They have no right angles; the large blue shape does not have 4 congruent sides) *What is it about the large orange shape that does not fit this description?* (It does not have 4 congruent sides.)

Remind students that as they sorted shapes they took shapes from Column 2 and put them in Columns 3, 4, and 5. This means that rectangles, squares, and rhombuses are all parallelograms. Have students look at their worksheets.

◆ **Say:** *When is a parallelogram also a rectangle?* (when it has 4 right angles) *When is a rhombus also a square?* (when it has 4 right angles) *When is a rectangle also a square?* (when it has 4 equal sides)

ACTIVITY THREE: Practice Drawing and Classifying Quadrilaterals

Materials: dot paper (BLM 13), triangular dot paper (BLM 12), centimeter ruler for each student

Draw a rhombus on a triangular dot paper transparency, as shown. Display the transparency on the overhead. Have students copy the rhombus on their triangular dot paper. Remind students that to classify quadrilaterals it helps to look for parallel sides, congruent sides, and right angles.

◆ **Say:** *How many pairs of parallel sides are there?* (2) *How many congruent sides are there?* (4) *Are there any right angles?* (No) *Is this a parallelogram?* (Yes) *Is this parallelogram also a rectangle, a square,* or a rhombus? (a rhombus) *Why is it not also a rectangle or a square?* (It does not have 4 right angles.)

Have students draw the following quadrilaterals:

• a quadrilateral with 2 pairs of parallel congruent sides and all right angles;

• a quadrilateral with 2 pairs of parallel congruent sides and no right angles;

• a quadrilateral with 4 congruent sides and 4 right angles;

• a quadrilateral with 4 congruent sides and no right angles.

Next to each quadrilateral have students write any names that apply.

SUMMING-IT-UP QUIZ

Give this quiz to students.

On the board, draw a square labeled *A*, a nonsquare rhombus labeled *B*, a trapezoid labeled *C*, and a rectangle labeled *D*.

◆ **Say:** *Which shape is a trapezoid?* (*C*) *How do you know?* (It has only one pair of parallel sides.)

◆ **Say:** *Which shape is a square?* (*A*) *Is square* A *also a rectangle? Why or why not?* (Yes. A rectangle is a shape that has 2 pairs of parallel congruent sides and 4 right angles. Square *A* has this, so it is also a rectangle.)

◆ **Say:** *Which shape is a rhombus with no right angles?* (*B*)

◆ **Say:** *Which shape is a rectangle, but is not also a square?* (*D*)

If students answer the *Summing-It-Up Quiz* correctly, they should then resume work on page 311 of their textbook. If students do not successfully complete the *Summing-It-Up Quiz*, further remediation may be necessary.

Formulas and Variables

Use with Lesson 10-9, exercises 1–10, text pages 348–349, and Lesson 10-12, exercises 1–11, text pages 354–355.

DIAGNOSTIC INTERVIEW

Before beginning this *Intervention Workshop* lesson, have students complete the following exercises.

Diagnostic Exercises

◆ **Say:** *Write the following. Tell which is an expression and which is a formula. Then name the variables and constants in each.*

$$2x + 2y \qquad P = 2l + 2w$$

($2x + 2y$ is an expression with variables x and y and constants 2 and 2; $P = 2l + 2w$ is a formula with variables P, l, w and constants 2 and 2.)

◆ **Say:** *Write the following. Evaluate each for* s = 12.

$$A = s \times s \qquad s/3 + 8 \qquad P = 4s$$

($A = 144$; 12; $P = 48$)

If students answer the Diagnostic Exercises correctly, there is probably no need for them to do this *Intervention Workshop* lesson. But before they resume work on page 349 or 355, you may want to assign the review lesson on page 28 of *Skills Update* in the student textbook. Then have students rejoin their class on page 349 or 355. For students who have had difficulty, continue this *Intervention Workshop*.

BACKGROUND

This *Intervention Workshop* lesson addresses the problem of the difference between constants and variables and how to evaluate expressions and formulas. Students begin with identifying constants and variables, and then evaluate expressions for given values using tiles. Next students work with formulas in the same manner. Being able to distinguish constants from variables is a necessary skill for students when they choose a formula and evaluate it and when they solve equations.

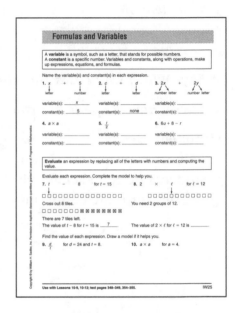

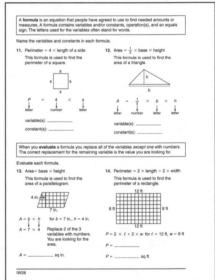

TEACHING THE *INTERVENTION WORKSHOP* LESSON

Getting Started

Materials: counters, centimeter grid paper (BLM 1)

◆ **Say:** *Write these expressions, compute, and write your answer with the correct units.*

      6 cm + 10 cm   4 × 8 in.   60 miles per hour × 2 hours

(16 cm; 32 in.; 120 miles)

◆ **Say:** *Write these expressions, compute, and write your answer with the correct units.*

      3 in. × 8 in.   5 m × 5 m

(24 in.^{2}; 25 m^2)

🕐 USING PAGES IW25 and IW26

Exercises 1–6 on page IW25
Have a student read aloud the definitions in the box at the top of page IW25. Then have a student read aloud the direction line above exercises 1-6.

Assist students as necessary to complete these exercises.

◆ **Say:** *Some of the operations you can perform on numbers are addition, subtraction, multiplication, and division. Each of these operations has a symbol. Write the symbol for each.* (+, −, ×, and ÷)

◆ **Say:** *Look at exercise 1. What operation do you see in this expression?* (addition)

You can have students identify the operation(s) in exercises 2-6 in the same way.

Exercises 7–10 on page IW25
Have a student read aloud the sentence in the box in the middle of page IW25. Next have a student read aloud the direction line above exercises 7-8. In these exercises the tile models are provided for students to interpret.

Assist students as they work on exercises 7 and 8.

◆ **Say:** *In exercise 7 count the cross-outs to be sure that 8 tiles are crossed out to show subtraction.*

◆ **Say:** *In exercise 8 remember that 2 times 12 means 2 groups of 12.*

Next have a student read aloud the direction line above exercises 9–10. In these exercises students may draw their own tile models to evaluate the expressions.

Guide students to complete these exercises. Suggest in exercise 9 that they draw 24 tiles and then ring 8 equal groups. For exercise 10 you can remind students that 4 × 4 means 4 groups of 4.

◆ **Say:** *You can use counters to help you evaluate expressions. You can move the counters around to form equal groups or move them aside to show subtraction. In this way you will see the correct value of the expression given.*

By identifying constants and variables and then evaluating expressions with models, exercises 1-10 provide students with a visual methodology to help them understand how variables and constants are used in expressions.

Exercises 11–12 on page IW26
Have a student read aloud the material in the box at the top of page IW26. Then have a student read aloud the direction line above exercises 11–12.

◆ **Say:** *These exercises are just like the ones you did on page 25, except now you are working with formulas instead of expressions.*

Assist students as necessary to complete exercises 11-12.

Next have a student read aloud the material in the box in the middle of page IW26. Then have a student read aloud the direction line above exercises 13-14.

Before beginning these exercises refer students to the exercises they completed earlier in *Getting Started* in the following way.

◆ **Say:** *Look back at your work in the* Getting Started *activity. How did you decide whether the units are square units?* (When you multiply units by units, the answer is in square units.)

SUMMING-IT-UP QUIZ

Give this quiz to students.

◆ **Say:** *Write the following expressions. Evaluate each for* h = 2 *and* t = 1.

    4h − 2t   *h × t*   *7h/t*   (6; 2; 14)

◆ **Say:** *Write the following formulas. Evaluate exercise* **a** *for* A = 24 m² *and* b = 6 m *and exercise* **b** *for* r = 45 miles per hour *and* t = 3 hours.

    a. *A = b × h*   ***b.*** *d = r × t*   (**a:** *h* = 4 m; **b:** *d* = 135 mi)

◆ **Say:** *Explain the difference between a variable and a constant.* (A variable stands for possible numbers while a constant is a specific number.)

If students answer the *Summing-It-Up Quiz* correctly, you may want to assign the review lesson on page 28 of *Skills Update* in the student textbook. Then students should return to Lesson 10-9 or 10-12 and resume work on page 349 or 355 of their textbook. If students do not successfully complete the *Summing-It-Up Quiz*, further remediation may be necessary.

Answers: Pages IW25 and IW26

2. *c*; *d*   **3.** variables: *x*; *y*; constants: 2; 2

4. variable: *a*; constant: none   **5.** variable: *t*; constant: 1/2

6. variables: *u, r*; constants: 6; 8

8. 24   **9.** 3   **10.** 16

11. variables: *P, s*; constant: 4   **12.** variables: *A, b, h*; constant: 1/2

13. 28   **14.** 2 × 12 + 2 × 8; 40

Formulas with Exponents

Extension lesson for the gifted and talented

Use with Lesson 10-9, exercises 1–10, text pages 348–349, and
Lesson 10-12, exercises 1–11, text pages 354–355.

BACKGROUND

Exercises 1–10 on page 349 of the student textbook require students
to use and write formulas for the perimeter of different polygons.
Exercises 1–11 on page 355 require students to use formulas for the
area of triangles and parallelograms. These *Intervention Workshop*
activities provide an opportunity to extend students' skills in writing
and using formulas by introducing students to formulas that involve
exponents. In Activity One students write a formula for the volume of
a cone, then use this formula to calculate volume. In Activity Two
students write and use a formula for finding the surface area of a cube.
In Activity Three students write and use a formula for calculating the
distance fallen in a given time by an object that is dropped from a height.
Use one or more of these activities with your students, as needed.

TEACHING THE *INTERVENTION WORKSHOP* LESSON

Getting Started

Write the following text on the board.

English Expression	Algebraic Expression
a number (n) squared plus 40	$n^2 + 40$
the product of a number (n) and 3	_____
four times the square of a number (n)	_____

Point out how the parts of the first English expression, *a number* (n)
squared plus 40, correspond to the algebraic expression $n^2 + 40$.
Then have students write an algebraic expression for each of the
other two English expressions on the board.

◆ **Say:** *What algebraic expression did you write for* the product of a
number (*n*) and 3? ($n \times 3$, or $3n$) *What algebraic expression did you
write for* four times the square of a number (n)? ($4 \times n^2$, or $4n^2$)

ACTIVITY ONE: Use a Formula to Find the Volume of a Sphere

Materials: ball or model of a sphere

Display this figure and text on the board.

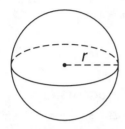

The volume of a sphere (V) is equal to 4/3 times π times the cube of the radius (r) of the sphere.

Hold up a ball, or a model of a sphere. Tell students that the sentence on the board describes how to find the volume of a sphere. Read the sentence aloud, then lead students through the process of writing a formula for the volume of a sphere.

◆ **Say:** *What operation or operations do you use to find the volume of a sphere?* (multiplication) *What amounts do you need to multiply in order to find the volume of a sphere?* (4/3, π, and the cube of the radius of a sphere)

◆ **Say:** *When a number is cubed, what number do we write for the exponent?* (3) *Let V represent the volume of a sphere and r represent the radius of the sphere. Write a formula that finds the volume of a sphere.* ($V = 4/3 \times \pi \times r^3$, or $4/3\pi r^3$)

Have a volunteer write the formula on the board. Then have students use the formula to find the volume of a sphere for the following values: $r = 5$ m, $r = 12$ m. Explain that students are to use 22/7 as an approximate value for π. Remind students that volume is expressed in cubic units. (523 17/21 m³; 7241 1/7 m³)

ACTIVITY TWO: Use a Formula to Find the Surface Area of a Cube

Display this figure and text on the board.

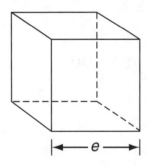

The surface area of a cube (SA) is equal to 6 times the square of the length of an edge (e).

Hold up a cube. Tell students that the sentence on the board describes how to find the surface area of a cube. Remind students that the surface area of a figure is the number of square units that exactly cover its six faces. Read the sentence on the board aloud, then lead students through the process of writing a formula for the surface area of a cube.

◆ **Say:** *What operation or operations do you use to find the surface area of a cube?* (multiplication) *What amounts do you need to multiply in order to find the surface area of a cube?* (6 and the square of the length of the edge of the cube)

◆ **Say:** *When a number is squared, what number do we write for the exponent?* (2) *Let SA represent the surface area of a cube and* e *represent the length of an edge. Write a formula that finds the surface area of a cube.* ($SA = 6 \times e^2$, or $SA = 6e^2$)

Have a volunteer write the formula on the board. Then have students use the formula to find the surface area of a cube for the following surface area: $e = 4$ in., $e = 5$ ft. Remind students that surface area is expressed in square units. (96 in.^{2}; 150 ft^2)

ACTIVITY THREE: Use the Vertical Motion Formula to Find the Distance Traveled Due to Gravity

Write these statements on the board.

The distance in feet (d) fallen due to Earth's gravitation is equal to 16 times the square of the time (t) that the object has been falling.

The time (t) is expressed in number of seconds.

Have students imagine dropping an object off a high cliff. Explain that the first sentence on the board tells how to find the number of feet the object has fallen after a given number of seconds. Read the sentence aloud, then lead students through the process of writing a formula for the distance fallen by an object due to gravity after a given time.

◆ **Say:** *What operation or operations do you use to find the distance an object falls?* (multiplication) *What amounts do you need to multiply in order to find the distance?* (16 and the square of the time that an object has been falling)

◆ **Say:** *When a number is squared, what number do we write for the exponent?* (2) *Let* d *represent the distance traveled by a falling object and* t *represent the time, in number of seconds, that the object has been falling. Write a formula that finds the distance fallen by a dropped object after a given number of seconds.* ($d = 16 \times t^2$, or $d = 16t^2$)

Have a volunteer write the formula on the board. Then have students use the formula to solve the following problems:

- *If you drop a penny into a well, and you hear it hit the bottom 1.5 seconds later, how deep is the well?* (36 ft)

- *If you drop a penny into a well, and you hear it hit the bottom 2 seconds later, how deep is the well?* (64 ft)

Equivalent Fractions

Use with Lesson 11-2, exercises 13–24, text pages 378–379, and
Lesson 11-4, exercises 1–8, text pages 382–383.

DIAGNOSTIC INTERVIEW

Before beginning this *Intervention Workshop* lesson, have students
complete the following exercises.

Diagnostic Exercises

Write this exercise on the board.

$$2/3 = ?/9$$

Have students work independently to complete the equivalent
fraction by finding the missing number.

◆ **Say:** *What is the missing number?* (6) *How did you find the missing
number?* (Possible response: To find equivalent fractions, multiply or
divide the numerator and the denominator by the same number. To
change a denominator from 3 to 9, multiply the denominator by 3;
so to find the numerator, multiply the numerator by 3.)

Write the following on the board: 2/8 and 4/12. Have students work
independently to decide if the fractions are equivalent.

◆ **Say:** *Are the fractions equivalent?* (No) *How do you know?*
(Possible response: If you write each fraction in simplest form,
2/8 = 1/4 and 4/12 = 1/3. 1/4 ≠ 1/3)

If students answer the Diagnostic Exercises correctly, there is probably no need
for them to do this *Intervention Workshop* lesson. Have those students rejoin
their class on page 379 or 383 of the student textbook. For students who have had
difficulties answering the exercises above, continue the *Intervention Workshop.*

BACKGROUND

Exercises 13–24 on page 379 of the student textbook require students to use their
understanding of equivalent fractions to find equal ratios. Exercises 1–8 on page 383 of
the student textbook require students to use equivalent fractions or the cross-products
rule to determine if two ratios form a proportion. These *Intervention Workshop*
activities address the needs of students who have not yet assimilated the skill of finding
equivalent fractions. In Activity One students use models to find equivalent fractions. In
Activity Two students review a computational process for finding equivalent fractions.
Activity Three provides an opportunity for students to decide if two fractions are
equivalent. Use one or more of these activities with your students, as needed.

TEACHING THE *INTERVENTION WORKSHOP* LESSON

Getting Started

Draw this fraction circle on the board.

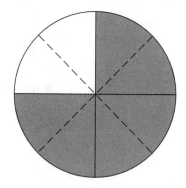

◆ **Say:** *How many fourths are shaded?* (3) *How many eighths are shaded?* (6) *Do 3/4 equal 6/8?* (Yes)

ACTIVITY ONE: Use Models to Find Equivalent Fractions

Materials: fraction strips (BLM 9)

Write the following on the board.

$2/5 = ?/10$

◆ **Say:** *What do you need to find out in order to solve the problem?* (how many tenths are equal to two fifths)

Tell students they will now use fraction strips to solve the problem. Have students find fraction strips for fifths and shade sections to show 2/5. Then have them find fraction strips for tenths and use the tenths to cover 2/5.

◆ **Say:** *How many tenths does it take to cover 2/5?* (4) *How many tenths are equal to 2/5?* (4)

On the board, complete the example so that it reads 2/5 = 4/10.

Have students use fraction strips to complete the following exercises.

$2/3 = ?/12$ (8)

$3/9 = ?/3$ (1)

$2/4 = ?/8$ (4)

$10/12 = ?/6$ (5)

ACTIVITY TWO: Use Computation to Find Equivalent Fractions

Write the following on the board.

$5/6 = ?/12$

◆ **Say:** *What do you need to find out in order to solve the problem?* (how many twelfths are equal to 5/6)

Lead students through the process of finding how many twelfths are equal to 5/6. Remind students that when they find equivalent fractions, they multiply or divide both the numerator and denominator by the same number.

◆ **Say:** *By what number was the original denominator, 6, multiplied to get the new denominator, 12?* (2) *Since the original denominator was multiplied by 2, the original numerator must also be multiplied by 2. What is the original numerator?* (5) *What is 5 times 2?* (10) *How many twelfths are equal to 5/6?* (10)

On the board, complete the example so that it reads 5/6 = 10/12.

Write the following on the board.

12/16 = 3/?

Lead students through the process of completing the equivalent fraction. Remind students that when they find equivalent fractions, they multiply or divide both the numerator and denominator by the same number.

◆ **Say:** *By what number was the original numerator, 12, divided to get the new numerator, 3?* (4) *Since the original numerator was divided by 4, the original denominator must also be divided by 4. What is the original denominator?* (16) *What is 16 divided by 4?* (4) *What is the new denominator?* (4)

On the board, complete the example so that it reads 12/16 = 3/4.

Have students use multiplication or division to complete each of the following.

3/5 = ?/10 (6)

7/8 = 21/? (24)

24/30 = ?/5 (4)

8/28 = 2/? (7)

ACTIVITY THREE: Use Models or Rename Fractions to Decide if Fractions Are Equivalent

Materials: fraction strips (BLM 9)

Write the following on the board.

6/8   8/12

◆ **Say:** *How could you use fraction strips to decide if 6/8 and 8/12 are equivalent fractions?* (Possible response: Shade fraction strips to show 6/8. Shade fraction strips to show 8/12. Align them to see if they are the same length.)

Have students use their fraction strips to decide if 6/8 and 8/12 are equivalent fractions.

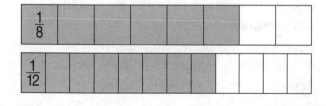

◆ **Say:** *Are 6/8 and 8/12 equivalent fractions?* (No)

Write the following on the board.

4/6   6/9

Tell students that another way of deciding if two fractions are equivalent is to write them both in lowest terms and see if they are the same.

◆ **Say:** *To write 4/6 in lowest terms, first find the greatest common factor of 4 and 6. Write the factors of 4 and 6. What is the greatest common factor of 4 and 6?* (2) *Divide the numerator and the denominator of 4/6 by the greatest common factor, 2. What is 4/6 in lowest terms?* (2/3) *Now find the greatest common factor of 6 and 9 to write 6/9 in lowest terms. Write the factors of 6 and 9. What is the greatest* *common factor of 6 and 9? (3) Divide the numerator and the denominator of 6/9 by the greatest common factor, 3. What is 6/9 in lowest terms? (2/3) Are 4/6 and 6/9 equivalent fractions?* (Yes)

Have students work independently to decide if the fractions in each of the following pairs are equivalent.

3/6	1/2 (Yes)
3/4	10/12 (No)
9/15	2/5 (No)
4/9	8/18 (Yes)
2/3	9/12 (No)

SUMMING-IT-UP QUIZ

Give this quiz to students.

Write this exercise on the board.

15/18 = ?/6

Have students work independently to complete the equivalent fraction by finding the missing number

◆ **Say:** *What is the missing number? (5) How did you find the missing number?* (Possible response: Divide the numerator and the denominator of the first fraction by the same number. To change a denominator from 18 to 6, divide the denominator by 3; so to find the numerator, divide the numerator by 3.)

Write the following on the board.

20/32   10/18

Have students work independently to decide if the fractions are equivalent.

◆ **Say:** *Are the fractions equivalent?* (No) *How do you know?* (Possible response: If you write each fraction in lowest terms, 20/32 = 5/8 and 10/14 = 5/9. 5/8 ≠ 5/9)

If students answer the *Summing-It-Up Quiz* correctly, they should then resume work on page 379 or 383 of their textbook. If students do not successfully complete the *Summing-It-Up Quiz*, further remediation may be necessary.

Solve an Equation Using Related Facts

Use with Lesson 11-5, exercises 3–14, text pages 384–385.

DIAGNOSTIC INTERVIEW

Before beginning this *Intervention Workshop* lesson, have students complete the following exercises.

Diagnostic Exercises

Write this multiplication sentence on the board.

$$n \times 20 = 100$$

◆ **Say:** *What related division sentences can you write using the variable* n *and the numbers 20 and 100?* $(100 \div 20 = n; 100 \div n = 20)$

Write this multiplication sentence on the board: n × *5 = 60.*

◆ **Say:** *What division sentence can help you solve this multiplication sentence?* (possible answer: $60 \div 5 = n$) *Complete the division. What value for* n *makes both the multiplication sentence and the division sentence true?* (12)

If students answer the Diagnostic Exercises correctly, there is probably no need for them to do this *Intervention Workshop* lesson. Have those students rejoin their class on page 385 of the student textbook. For students who have had difficulties answering the exercises above, continue the *Intervention Workshop*.

BACKGROUND

Exercises 3–14 on page 385 of the student textbook require students to use related division sentences in solving proportions. These *Intervention Workshop* activities address the needs of students who have difficulty identifying related division sentences and using those sentences to solve multiplication equations. In Activity One students use models to help them find and solve related multiplication and division equations. In Activity Two students use the meanings of multiplication and division to help them find and solve related multiplication and division equations. In Activity Three students practice finding and solving related multiplication and division equations. Use one or more of these activities with your students, as needed.

TEACHING THE *INTERVENTION WORKSHOP* LESSON

Getting Started

Write the following on the board.

$n \times 4 = 24$

◆ **Say:** *What number times 4 equals 24?* (6) *What is a related division fact for 6 times 4 equals 24?* ($24 \div 4 = 6$; $24 \div 6 = 4$)

ACTIVITY ONE: Use Models to Find and Solve Related Multiplication and Division Equations

Materials: base ten blocks

Write the following on the board.

$n \times 30 = 180$

◆ **Say:** *How could you use base ten blocks to solve the equation?* (Possible responses: Make groups of 30 and find how many groups of 30 it takes to make 180; show 180, then separate 180 into groups of 30 and count the number of groups you have.)

Have students show 180, then separate 180 into groups of 30. Remind them to trade 1 hundred flat for 10 ten rods.

◆ **Say:** *How many groups of 30 are in 180?* (6) *What division sentence could you write to represent separating 180 into groups of 30?* ($180 \div 30 = 6$) *What is the value of* n *in the multiplication sentence on the board?* (6)

Have students work in pairs to solve multiplication equations such as $n \times 21 = 84$, $n \times 40 = 120$, and $n \times 6 = 72$. Students begin by identifying the number in each group and the total. Students model the total and separate the total into equal groups. Students then write the corresponding division sentence and use that answer to find the value for *n* in the multiplication sentence. (4; 3; 12)

ACTIVITY TWO: Use the Meaning of Multiplication and the Meaning of Division to Find and Solve Related Equations

Materials: multiplication worksheet (BLM 15), division worksheet (BLM 16) for each student

Write the multiplication sentence $n \times 6 = 150$ on the board. Then have students write the sentence in one row of a multiplication worksheet.

Number of Sets	×	Number in Each Set	=	Number in All
n	×	6	=	150
	×		=	
	×		=	

◆ **Say:** *What is the total number in this multiplication example?* (150) *What is the number in each set?* (6) *What do you need to find?* (how many sets of 6 are in 150) *How could you use division to find how many sets of 6 are in 150?* (Divide 150 by 6.)

Have students write the numbers and the variable from the multiplication worksheet in the appropriate columns in a division worksheet.

Number in All	÷	Number in Each Set	=	Number of Sets
150	÷	6	=	n
	÷		=	
	÷		=	

◆ **Say:** *What division sentence can help you solve the multiplication sentence* n *times 6 equals 150?* (150 ÷ 6 = n) *Complete the division. What does* n *equal?* (25)

Have students work in pairs to solve multiplication equations such as $n \times 20 = 100$, $n \times 4 = 92$, and $n \times 14 = 70$. Tell students to begin by writing each example in the multiplication worksheet. Students then write the same data in the division worksheet, solve the division exercise, and find the value of n. (5; 23; 5)

ACTIVITY THREE: Practice Finding and Solving Related Multiplication and Division Equations

Materials: spinner divided into six equal sections with the numbers 4, 8, 12, 16, 24, and 48; number cards for 96, 144, 192, 240, 288, and 336

Write the following on the board.

$$n \times 6 = 78$$

◆ **Say:** *What division sentence can help you find the value of* n *in this multiplication sentence?* (possible response: 78 ÷ 6 = n) *Complete the division. What value for n makes both the multiplication sentence and the division sentence true?* (13)

Have students work in pairs. Explain that one partner uses the spinner and the number cards to write a multiplication sentence. The first factor in the multiplication sentence is n. The second factor is found by spinning the spinner. The product is found by picking a number card. For example if one student spins an 8 and picks a card for 144, he/she writes the multiplication sentence $n \times 8 = 144$.

The partner then writes the related division sentence, 144 ÷ 8 = n. The first student finds the value for n that makes both the multiplication sentence and division sentence true. Partners then switch roles and repeat the activity. Students continue until they have solved 6 different multiplication equations.

SUMMING-IT-UP QUIZ

Give this quiz to students.

Write this multiplication sentence on the board.

$$n \times 5 = 135$$

◆ **Say:** *What related division sentences can you write using the variable* n *and the numbers 5 and 135?* ($135 \div 5 = n$; $135 \div n = 5$)

Write this multiplication sentence on the board.

$$n \times 16 = 96$$

◆ **Say:** *What division sentence can help you solve this multiplication sentence?* (possible answer: $96 \div 16 = n$) *Complete the division. What value for* n *makes both the multiplication sentence and the division sentence true?* (6)

If students answer the *Summing-It-Up Quiz* correctly, they should then resume work on page 385 of their textbook. If students do not successfully complete the *Summing-It-Up Quiz*, further remediation may be necessary.

DIAGNOSTIC INTERVIEW

Before beginning this *Intervention Workshop* lesson, have students complete the following exercises.

Diagnostic Exercises

◆ **Say:** *Write these percents. Write each as a decimal and a fraction.*

      25%   50%   75%

(0.25, 1/4; 0.5, 1/2; 0.75, 3/4)

◆ **Say:** *Write the following. Find the percent of the number.*

      25% of 76   50% of 130   75% of 64

(19; 65; 48)

If students answer the Diagnostic Exercises correctly, there is probably no need for them to do this *Intervention Workshop* lesson. Have those students rejoin their class on page 415 of the student textbook. For students who have had difficulty answering the exercises above, continue the *Intervention Workshop*.

BACKGROUND

This *Intervention Workshop* lesson addresses the problem of the relationship between fractions, decimals, and percents and how to use this relationship to find the percent of a number. Students begin with identifying common percents and their decimal and fraction equivalents using a variety of models. Next students develop their percent/fraction sense by using rectangles and line segments to find the percent of a number. These models provide a clear visual methodology for students to find the percent of a number mentally and to check the reasonableness of their answers in solving percent problems.

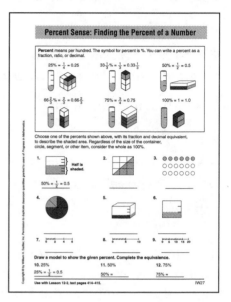

TEACHING THE *INTERVENTION WORKSHOP* LESSON

Getting Started

Materials: centimeter grid paper (BLM 1)

◆ **Say:** *Write these fractions. Then write each as a decimal.*

     5/100   *1/5*   *35/100*

(0.05; 0.2; 0.35)

◆ **Say:** *Draw 12 circles. Shade 1/2 of them. Cross out 3/4 of them. Tell how many circles are unshaded.* (6) *Tell how many are crossed out.* (9)

🕐 USING PAGES IW27 and IW28

Exercises 1–12 on page IW27
Have a student read aloud the material in the box at the top of page IW27. Then have a student read aloud the direction line above exercises 1–9. Assist students as necessary to complete these exercises.

Have a student read aloud the direction line above exercises 10–12. Guide students as necessary to complete these exercises.

◆ **Say:** *You can make models by using grid paper and drawing squares or rectangles as was done in exercise 2. Shade the appropriate number of squares to show the percent.*

By developing the relationship between common percents and their fraction and decimal equivalents, exercises 1–12 provide students with a visual methodology to help them develop their percent/fraction sense.

Exercises 13–18 on page IW28
Have a student read aloud the material in the box at the top of page IW28. Then have a student read aloud the direction line in exercise 13.

◆ **Say:** *In exercises 13-15 the rectangles are shaded for you. The number of shaded small squares shows the fraction equivalent for the percent. You can look at page IW27 to check the fraction equivalent for the percent.*

Assist students as necessary to complete exercises 13-15.

Have a student read aloud the direction line above exercises 16-18. In these exercises students shade the appropriate number of small squares to find the percent of the number.

Exercises 19–21 on page IW28
Have a student read aloud the direction line above exercises 19-21. Assist students as necessary to complete these exercises.

◆ **Say:** *If you need help with these exercises, look at exercises 7-9 on page IW27. These exercises are similar.*

By using rectangles and line segment models, exercises 13-21 guide students to find the percent of a number.

SUMMING-IT-UP QUIZ

Give this quiz to students.

◆ **Say:** *Write these percents. Then write each as a decimal and a fraction.*

$$33 \, 1/3\% \qquad 100\% \qquad 66 \, 2/3\%$$

(0.33 1/3 = 1/3; 1.0 = 1/1; 0.66 2/3 = 2/3)

◆ **Say:** *Write the following. Then find the percent of the number.*

$$33 \, 1/3\% \text{ of } 57 \qquad 100\% \text{ of } 100 \qquad 66 \, 2/3\% \text{ of } 600$$

(19; 100; 400)

If students answer the *Summing-It-Up Quiz* correctly, then they should return to Lesson 12-2 and resume work on page 415 of their textbook. If students do not successfully complete the *Summing-It-Up Quiz*, further remediation may be necessary.

Answers: Pages IW27 and IW28

2. 50% = 1/2 = 0.5   **3.** 33 1/3% = 1/3 = 0.33 1/3

4. 75% = 3/4 = 0.75   **5.** 25% = 1/4 = 0.25

6. 33 1/3 % = 1/3 = 0.33 1/3   **7.** 33 1/3% = 1/3 = 0.33 1/3

8. 50% = 1/2 = 0.5   **9.** 75% = 3/4 = 0.75

11. 1/2 = 0.5   **12.** 3/4 = 0.75   **14.** 5

15. 1   **16.** 4   **17.** 8

18. 1   **20.** 4; 4   **21.** 5; 5

Using Percent

Use with Lesson 12-5, exercises 1–10, text pages 420–421.

DIAGNOSTIC INTERVIEW

Before beginning this *Intervention Workshop* lesson, have students complete the following exercises.

Diagnostic Exercises

Write the following on the board.

> 47% of 80 → about 1/4, 1/3, or 1/2 of 80?

◆ **Say:** *How would you estimate 47 percent of 80: by finding 1/4 of 80, 1/3 of 80, or 1/2 of 80?* (1/2 of 80) *Why?* (47% is close to 50%, which is equal to 1/2.) *Estimate 47 percent of 80. What is your estimate?* (40)

Write the following on the board.

> 7 = n% of 20

◆ **Say:** *How can you use fractions to find the value of* n*?* (Write the fraction 7/20. Rename the fraction as a decimal. Rename the decimal as a percent.) *Find the percent. What percent of 20 is 7?* (35%)

If students answer the Diagnostic Exercises correctly, there is probably no need for them to do this *Intervention Workshop* lesson. Have those students rejoin their class on page 421 of the student textbook. For students who have had difficulties answering the exercises above, continue the *Intervention Workshop*.

BACKGROUND

Exercises 1–10 on page 421 of the student textbook require students to solve problems by estimating and finding a percent of a number or by estimating and finding what percent one number is of another. These *Intervention Workshop* activities address the needs of students who have difficulty applying the concept of percent. Activity One provides students with an opportunity to find fractions for benchmark percents such as 10%, 25%, and 50%; understanding these equivalents will help students as they estimate and find percents of numbers. In Activity Two students use percent-fraction equivalents to estimate the percent of a number. In Activity Three students use the meaning of percents to help them estimate what percent one number is of another. In Activity Four students identify the components of a percent problem: rate (r), base (b), and percentage

(p), and use the equation $r \times b = p$ to help them solve problems involving percent. Use one or more of these activities with your students, as needed.

TEACHING THE *INTERVENTION WORKSHOP* LESSON

Getting Started

◆ **Say:** *You can rename a percent as a decimal and a fraction. Write a decimal that is equal to 25 percent.* (0.25) *Write a fraction in lowest terms that is equal to 25 percent.* (1/4)

If necessary, give students other percents to rename as decimals and as fractions in lowest terms.

ACTIVITY ONE: Find Percent-Fraction Equivalents

Materials: percent-fraction equivalence worksheet (BLM 17)

Remind students that *percent* means the *part of each hundred*. Then lead them to find the fractional equivalents of benchmark percents as indicated at the right. As students find each equivalent, have them list the percent and the equivalent fraction in a table, as shown.

Percent	Fraction
25%	$\frac{1}{4}$
50%	$\frac{2}{4}$ or $\frac{1}{2}$
75%	$\frac{3}{4}$
10%	
20%	
30%	
40%	
60%	
70%	
80%	
90%	
$33\frac{1}{3}$%	
$66\frac{2}{3}$%	

◆ **Say:** *What fraction with a denominator of 100 can you write for 25%?* (25/100) *What is that fraction in lowest terms?* (1/4) *What fraction in lowest terms can you write for 50%?* (1/2) *What fraction in lowest terms can you write for 75%?* (3/4)

Continue by having students find fractional equivalents for 10%, 30%, 40%, 60%, 70%, 80% and 90%, and listing those equivalents in the table. (1/10; 3/10; 2/5; 3/5; 7/10; 4/5; 9/10) Finally have students list any other commonly used percents and fractional equivalents they know, such as 33 1/3% and 1/3, and 66 2/3% and 2/3.

ACTIVITY TWO: Use Percent-Fraction Equivalents to Estimate a Percent of a Number

Materials: percent-fraction equivalence worksheet (BLM 17)

Write the following on the board.

Estimate 26% of 79

Explain that students can use fractions to estimate a percent of a number. If students completed *Activity One*, have them take out their percent-fraction equivalence tables. Then lead students through the estimation process.

◆ **Say:** *What percent is close to 26% and is easy to compute with mentally?* (25%) *What fraction can you use for that percent?* (1/4)

◆ **Say:** *What is a number that is close to 79 and easy to divide into fourths, that is, divide by 4?* (80) *What is 1/4 of 80?* (20)

◆ **Say:** *So 26 percent of 79 is about 20.*

Lead students through another example of the estimation process. Write the following on the board.

> *Estimate 8% of 124.*

◆ **Say:** *What percent is close to 8% and is easy to compute with mentally?* (10%) *What fraction can you use for that percent?* (1/10)

◆ **Say:** *What is a number that is close to 124 and easy to divide into tenths?* (120) *What is 1/10 of 120?* (12)

◆ **Say:** *So 8 percent of 124 is about 12.*

Have students work independently to estimate each of the following. Accept estimates that students can reasonably justify.

> 52% of 349 (175)
> 73% of 33 (24)
> 91% of 204 (180)
> 34% of 595 (200)

ACTIVITY THREE: Use the Meaning of Percents to Estimate What Percent One Number is of Another

Write the following on the board.

> *About what percent of 53 is 16?*

Explain that to make this estimate students can use the meaning of *percent*, which is *part of 100*. Write the following on the board.

> *16 of 53 = ? of 100; 16/53 = ?/100*

◆ **Say:** *What denominator is close to 53 and can easily be renamed as 100?* (50)

Write *16/50* on the board.

◆ **Say:** *How many hundredths are equal to 16/50?* (32) *About what percent of 53 is 16?* (32%)

Tell students that another way to estimate what percent one number is of another is to look for fractions for which percents are easy to find, such as 1/2, 1/3, 1/4, or 1/10.

Write the following on the board.

> *22 is about what percent of 84?*

◆ **Say:** *What fraction means 22 of 84?* (22/84) *What fraction is close to 22/84, but can be renamed as a fraction that is easy to work with, such as 1/2, 1/3, or 1/4?* (22/84 is close to 20/80, which is equivalent to 1/4.) *About what percent of 84 is 22?* (25%)

Have students work independently to estimate each of the following. Accept estimates that students can reasonably justify.

- About what percent of 26 is 14? (56%)
- About what percent of 201 is 94? (47%)
- About what percent of 59 is 41? (66 2/3%)
- About what percent of 12 is 35? (300%)

ACTIVITY FOUR: Identify the Rate, Base, and Percentage in Problems Involving Percent

Materials: rate, base, percentage worksheet (BLM 18), percent-fraction equivalence worksheet (BLM 17)

Write the following on the board.

What is 25% of 68?

Have students write the information from that question in their rate, base, and percentage worksheets.

Rate percent (%)	of (×)	Base total number	=	Percentage number that represents part of total
25	×	68	=	
	×		=	
	×		=	

◆ **Say:** *In this problem what is the rate, or percent?* (25%) *What fraction can you use for 25 percent?* (1/4) *What decimal can you use for 25 percent?* (0.25)

◆ **Say:** *What is the base, or total number?* (68)

◆ **Say:** *What do you need to find?* (the percentage) *How can you find the percentage?* (Possible responses: Multiply 1/4 × 68; multiply 0.25 × 68.)

◆ **Say:** *Find the percentage. What is 25 percent of 68?* (17)

Tell students that you will present a problem situation and that they will take notes on the data and write the data in their worksheets. Then present this problem:

• *In a survey of 570 students, 10% said they played tennis regularly. How many students said they played tennis regularly?*

◆ **Say:** *In this problem what is the rate, or percent?* (10%) *What fraction can you use*

for 10 percent? (1/10) *What decimal can you use for 10 percent?* (0.1)

◆ **Say:** *What is the base, or total number?* (570)

◆ **Say:** *What do you need to find?* (the percentage) *How can you find the percentage?* (Possible responses: Multiply 1/10 × 570; multiply 0.1 × 570.)

◆ **Say:** *Find the percentage. How many students said they played tennis regularly?* (57)

Tell students they will now use their worksheets to find what percent one number is of another. Write the following on the board.

What percent of 40 is 12?

Have students write the information from that question in their rate, base, and percentage worksheets.

Rate percent (%)	of (×)	Base total number	=	Percentage number that represents part of total
	×	40	=	12
	×		=	
	×		=	

◆ **Say:** *In this problem what is the base, or total number?* (40) *What is the percentage, or number that represents part of the total?* (12) *What do you need to find?* (the rate, or percent, that 12 is of 40)

Write the following on the board.

n × 40 = 12

◆ **Say:** *What related division sentence can you use to find the solution to the multiplication sentence on the board?*

(12 ÷ 40 = *n*) *Complete the division, giving your answer as a decimal. What decimal is equal to 12 divided by 40?* (0.3) *What percent is equal to 3/10?* (30%) *What percent of 40 is 12?* (30%)

Tell students that you will present a problem situation, and that they will take notes on the data and write the data in their worksheets. Then present this problem:

Six of the players on a sports team are injured. There are 30 players in all. What percent of the team is injured?

◆ **Say:** *In this problem what is the base, or total number?* (30) *What is the percentage,*

or number that represents part of the total? (6) *What do you need to find?* (the rate, or percent, that 6 is of 30)

Write the following on the board.

n × 30 = 6

◆ **Say:** *What related division sentence can you use to help you find the solution to the multiplication sentence on the board?* (6 ÷ 30 = *n*) *Complete the division, giving your answer as a decimal. What decimal is equal to 6 divided by 30?* (0.2) *What percent is equal to 2/10?* (20%) *What percent of the team is injured?* (20%)

SUMMING-IT-UP QUIZ

Give this quiz to students.

Write the following on the board.

  Estimate 76% of 39.

◆ **Say:** *What percent is close to 76 percent and is easy to compute with mentally?* (75%) *What fraction can you use for that percent?* (3/4) *What is a number close to 39 and easy to divide into fourths?* (40) *What is 3/4 of 40?* (30) *What is your estimate?* (30)

Write the following on the board.

  What is 30% of 70?

◆ **Say:** *In this problem what is the rate, or percent?* (30%) *What fraction can you use for 30 percent?* (3/10) *What decimal can you use for 30 percent?* (0.3)

◆ **Say:** *What is the base, or total number?* (70)

◆ **Say:** *What do you need to find?* (the percentage) *How can you find the percentage?* (Possible responses: Multiply 3/10 × 70; multiply 0.3 × 70.)

◆ **Say:** *Find the percentage. What is 30 percent of 70?* (21)

Write the following on the board.

  What percent of 40 is 20?

◆ **Say:** *How can you find the answer to this problem?* (Possible responses: Find a decimal for 20/40, then write the decimal as a percent; divide 20 by 40 and write a percent for the answer; find the answer mentally.) *What percent of 40 is 20?* (50%)

If students answer the *Summing-It-Up Quiz* correctly, they should then resume work on page 421 of their textbook. If students do not successfully complete the *Summing-It-Up Quiz,* further remediation may be necessary.

DIAGNOSTIC INTERVIEW

Before beginning this *Intervention Workshop* lesson, have students complete the following exercises.

Diagnostic Exercises

◆ **Say:** *Write the following addition exercises. Compute to find each sum.*

$$^+3 + {}^+5 \qquad {}^+5 + {}^-3 \qquad {}^-5 + {}^+3 \qquad {}^-5 + {}^-3$$

($^+8$; $^+2$; $^-2$; $^-8$)

◆ **Say:** *Write the following subtraction exercises. Compute to find each difference.*

$$^+5 - {}^+3 \qquad {}^+5 - {}^-3 \qquad {}^-5 - {}^+3 \qquad {}^-5 - {}^-3$$

($^+2$; $^+8$; $^-8$; $^-2$)

If students answer the Diagnostic Exercises correctly, there is probably no need for them to do this *Intervention Workshop* lesson. Have those students rejoin their class on page 453 of the student textbook. For students who have had difficulty answering the exercises above, continue the *Intervention Workshop*.

BACKGROUND

This *Intervention Workshop* lesson addresses adding and subtracting integers and how to determine the sign of the sum or difference. Students begin by using tiles to model adding and subtracting integers. Then they use number lines to add and subtract integers. By using these models students are provided with a visual methodology to find the sums and differences of integers and understand how to find the correct sign of their answers.

TEACHING THE *INTERVENTION WORKSHOP* LESSON

Getting Started

Materials: 2-color counters

◆ **Say:** *To add and subtract integers, it is helpful to think of positive integers as gains and negative integers as losses. For example, $^+5$ dollars is a gain of 5 dollars and $^-4$ dollars is a loss of 4 dollars.*

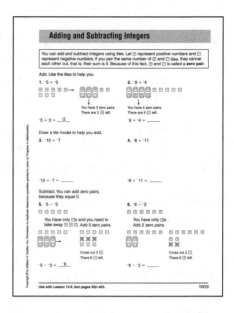

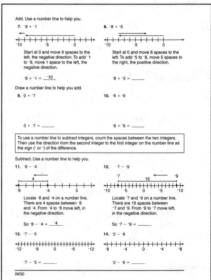

◆ **Say:** *Write these examples as gains and losses. Then write the sum. Is the sum a gain or a loss?*

$$^-6 + {}^-8 \qquad {}^+9 + {}^-12 \qquad {}^-10 + {}^+2$$

(loss of 6 plus loss of 8 = loss of 14; gain of 9 plus loss of 12 = loss of 3; loss of 10 plus gain of 2 = loss of 8)

◆ **Say:** *When you subtract integers, think of it as taking away a gain or loss. Write these examples as gains or losses, then write the difference. Is the difference a gain or a loss?*

$$^+8 - {}^+3 \qquad {}^-13 - {}^-2 \qquad {}^-6 - 0$$

(gain of 8 take away gain of 3 = gain of 5; loss of 13 take away loss of 2 = loss of 11; loss of 6 take away zero = loss of 6)

⏱ USING PAGES IW29 and IW30

Have a student read aloud the material in the box at the top of page IW29. Relate the definition of a zero pair to the explanation of gains and losses in *Getting Started* by drawing and explaining *zero pair* in the following way.

◆ **Say:** *This symbol or model* [draw +] *means* $^+1$ *or a gain of 1; this model* [draw −] *means* $^-1$ *or a loss of 1. The sum* $^+1 + {}^-1$ *equals zero just as a model of a zero pair* [draw + and −] *together equals zero.*

Exercises 1–6 on page IW29
Have a student read aloud the direction line above exercises 1-2. In these two exercises students are guided to find the sum of two integers with tiles.

Have a student read aloud the direction line above exercises 3-4.

◆ **Say:** *In these two exercises you draw your own tile model to add the integers. Remember to look for and ring zero pairs. It can be helpful to draw all "plus" tiles in one row and all "minus" tiles in a row just below.*

Have a student read aloud the direction line above exercises 5-6.

Assist students as necessary to complete these exercises in the following way.

◆ **Say:** *In exercise 5 you need to add 3 zero pairs so that you have* $^+3$ *to subtract. You cross out tiles to show the subtraction. Count the number of cross-outs to be sure that you have subtracted the correct number.*

If students have difficulty understanding adding zero pairs, remind them about the material in the box at the top of the page in the following way.

◆ **Say:** *Look back at the box at the top of this page. If you read the material again you will see that if you pair a "plus" tile and a "minus" tile, they cancel each other out. In other words, their sum is zero. This is why you can add any number of zero pairs to a subtraction problem and not change the difference.*

Exercises 1-6 guide students to add and subtract integers with tiles.

Exercises 7–14 on page IW30
◆ **Say:** *In the exercises on this page you will use a number line to add and subtract integers.*

Have a student read aloud the direction line above exercises 7–8.

◆ **Say:** *When you add integers on a number line, a move to the right shows a positive integer and a move to the left shows a negative integer.*

◆ **Say:** *After you locate the first integer in the addition, continue from that point, left or right, to show the second integer. Do not return to zero to show the second integer.*

Have a student read aloud the direction line above exercises 9–10. In these two exercises assist students as necessary to draw their own number line addition models.

Then have a student read aloud the material in the box in the middle of the page. Students may need guidance to understand the process of finding the correct sign of the difference.

Guide students to work through exercises 11–14 in the following way.

◆ **Say:** *Follow these steps.*

1. *On a number line graph each integer in the subtraction.*

2. *Count the number of spaces between the two integers to find the numerical (number) value of the answer.*

3. *Look at the second integer in the subtraction you graphed on a number line. To move from this integer to the first integer, do you move to the right or to the left?*

4. *If the direction is to the right, the sign of the answer is positive. If the direction is to the left, the sign of the answer is negative.*

Exercises 7–14 guide students to add and subtract integers using a number line.

SUMMING-IT-UP QUIZ

Give this quiz to students.

◆ **Say:** *Write the following addition exercises. Compute to find each sum.*

$$^+7 + {}^+2 \qquad ^+7 + {}^-2 \qquad ^-7 + {}^+2 \qquad ^-7 + {}^-2$$

($^+9$; $^+5$; $^-5$; $^-9$)

◆ **Say:** *Write the following subtraction exercises. Compute to find each difference.*

$$^+9 - {}^+2 \qquad ^+9 - {}^-2 \qquad ^-9 - {}^+2 \qquad ^-9 - {}^-2$$

($^+7$; $^+11$; $^-11$; $^-7$)

If students answer the *Summing-It-Up Quiz* correctly, they should return to Lesson 13-6 and resume work on page 453 of their textbook. If students do not successfully complete the *Summing-It-Up Quiz*, further remediation may be necessary.

Answers: Pages IW29 and IW30

2. $^-5$ **3.** $^-17$ **4.** $^-3$ **6.** $^+8$

8. $^-3$ **9.** $^-7$ **10.** 0 **12.** $^-16$

13. $^+12$ **14.** $^+8$

Locating Points on Horizontal and Vertical Number Lines

Use with Lesson 13-10, exercises 1–22, text pages 460–461.

DIAGNOSTIC INTERVIEW

Before beginning this *Intervention Workshop* lesson, have students complete the following exercises.

Diagnostic Exercises

On a transparency draw the coordinate graph shown. Display the transparency on the overhead. Have students copy the graph onto grid paper. Then write ($^-3$, $^+2$) below the graph.

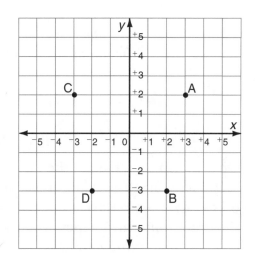

◆ **Say:** *What does the first number in the ordered pair negative 3, positive 2 [($^-3$, $^+2$)] tell you?* (It tells you that the point is 3 units to the left of zero.) *What does the second number in the ordered pair tell you?* (It tells you that the point is also two units above zero.) *What point is named by the ordered pair negative 3, positive 2?* (point C)

◆ **Say:** *What ordered pair can you write for point B?* [($^+2$, $^-3$)]

If students answer the Diagnostic Exercises correctly, there is probably no need for them to do this *Intervention Workshop* lesson. Have those students rejoin their class on page 461 of the student textbook. For students who have had difficulties answering the exercises above, continue the *Intervention Workshop*.

BACKGROUND

In exercises 1–22 on page 461 of the student textbook students graph ordered pairs of integers and identify the coordinates of points. These *Intervention Workshop* activities address the needs of students who have not yet mastered interpreting the meaning of ordered pairs and using a coordinate graph that has four quadrants. Activity One reviews locating and graphing ordered pairs in the first quadrant. In Activity Two students identify the meaning of each coordinate in ordered pairs involving all four quadrants. Activity Three provides students with an opportunity to practice identifying the coordinates of points. Use one or more of these activities with your students, as needed.

TEACHING THE *INTERVENTION WORKSHOP* LESSON

Getting Started

Draw a horizontal number line on the board. Mark points at ⁻1, 0, and ⁺1, only.

◆ **Say:** *Will 3 be to the left or the right of zero on the number line?* (right)
Will negative 5 be to the left or the right of zero on the number line? (left)

Draw a vertical number line on the board. Mark points at ⁻1, 0, and ⁺1, only.

◆ **Say:** *Will 2 be above or below zero on the number line?* (above)
Will negative 4 be above or below zero on the number line? (below).

ACTIVITY ONE: Graph Ordered Pairs in the First Quadrant

Materials: centimeter grid paper (BLM 1)

On a transparency draw the coordinate graph shown. Display the transparency on the overhead. Have students copy the graph onto grid paper. Then write (⁺4, ⁺3) below the graph.

◆ **Say:** *Put your finger at zero. Does the first number in the ordered pair tell you to move up, or to the right?* (to the right) *How many units to the right does the first number in the ordered pair tell you to move?* (4 units to the right) *Move your finger 4 units to the right.*

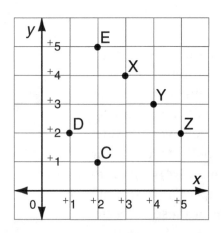

Point to zero and move your finger 4 units to the right along the horizontal axis.

◆ **Say:** *Does the second number in the ordered pair tell you to move up, or to the right?* (up) *How many units up does the second number in the ordered pair tell you to move?* (3 units up) *Start at the point 4 units to the right along the horizontal axis and move your finger 3 units up. What point does the ordered pair 4, 3 [(⁺4, ⁺3)] name?* (point Y)

Now have students write the ordered pair for a given point.

◆ **Say:** *Find point D. How many units to the right of zero is point D?* (1) *How many units do you move up to get to point D?* (2) *What ordered pair can you write for point D?* [(⁺1, ⁺2)]

Have students work independently to find the point that is located at (⁺5, ⁺2) and to write an ordered pair for point E. [Z; (⁺2, ⁺5)]

ACTIVITY TWO: Graph Ordered Pairs and Locate Points in Four Quadrants

Materials: centimeter grid paper (BLM 1)

On a transparency draw the chart and the coordinate graph shown. Display the transparency on the overhead. Have students copy the graph onto grid paper. Then write ($^-$3, $^+$5) below the graph.

(First Number x,	Second Number y)
positive (+) ⟶ ⟵ negative (−)	positive (+) ↑ negative(−) ↓

◆ **Say:** *In what direction does the first number in the ordered pair negative 3, positive 5 [($^-$3, $^+$5)] tell you to move?* (to the left) *How many units to the left does the first number in the ordered pair tell you to move?* (3 units to the left) *Start at zero and move your finger 3 units to the left.*

Point to zero and move your finger 3 units to the left along the horizontal axis.

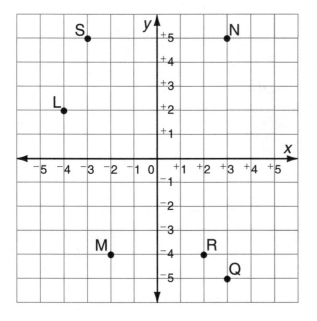

◆ **Say:** *In what direction does the second number in the ordered pair negative 3, positive 5 [($^-$3, $^+$5)] tell you to move?* (up) *How many units up does the second number in the ordered pair tell you to move?* (5 units up.) *Start at the point 3 units to the left along the horizontal axis and move your finger 5 units up. What point does the ordered pair negative 3, positive 5 [($^-$3, $^+$5)] name?* (S)

Now have students give the ordered pair for a given point.

◆ **Say:** *Find point R. If you start at zero, do you move right or left to get to point R?* (right) *How many units do you move right to get to point R?* (2) *When you move to the right of zero, are the numbers positive or negative?* (positive) *What is the x-coordinate of point R?* ($^+$2) *Do you move up or down to get to point R?* (down) *How many units do you move down?* (4) *When you move below zero, are the numbers positive or negative?* (negative) *What is the y-coordinate of point R?* ($^-$4) *What ordered pair can you write for point R?* [($^+$2, $^-$4)]

Have students work independently to find the point with coordinates ($^-$2, $^-$4) and to write an ordered pair for point Q. [M; ($^+$3, $^-$5)]

ACTIVITY THREE: Practice Graphing and Finding Coordinates of Points

Materials: centimeter grid paper (BLM 1)

Write the following on the board.

Point A ($^+4$, $^-1$)

◆ **Say:** *What does the first number tell you about point* A? (Point A is 4 units to the right of zero.) *What does the second number tell you about point* A? (Point A is also 1 unit below zero.)

Have students work in pairs. Each student draws a grid that goes from $^-10$ to $^+10$ on both axes, marks 4 points A, B, C, and D at any four intersections of grid lines, then connects the points to make a figure. Students exchange grids and write the coordinates for each point. Partners then check each others' work.

SUMMING-IT-UP QUIZ

Give this quiz to students.

On a transparency draw the coordinate graph shown. Display the transparency on the overhead. Have students copy the graph onto grid paper.

◆ **Say:** *What does the first number in any ordered pair tell you?* (whether to move right or left from 0) *What does the second number in any ordered pair tell you?* (whether to move up or down from 0)

Write ($^-5$, $^+1$) below the graph.

◆ **Say:** *What point does negative 5, positive 1 [($^-5$, $^+1$)] name?* (I) *What ordered pair can you write for point H?* [($^-1$, $^+5$)]

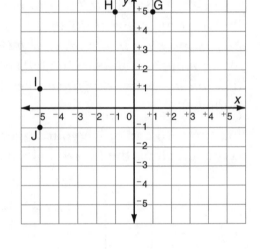

If students answer the *Summing-It-Up Quiz* correctly, they should then resume work on page 461 of their textbook. If students do not successfully complete the *Summing-It-Up Quiz*, further remediation may be necessary.

Meaning of Variables: Numerical and Algebraic Expressions

Use with Lesson 14-1, exercises 11–16, text pages 474–475.

DIAGNOSTIC INTERVIEW

Before beginning this *Intervention Workshop* lesson, have students complete the following exercises.

Diagnostic Exercises

Write the following on the board.

    1/2 of a number, minus 16

◆ **Say:** *Let the variable* n *stand for the unknown number. Translate the expression to an algebraic expression.* ($1/2 \cdot n - 16$, or $1/2\,n - 16$)

Write the following on the board.

    84 plus twice z

◆ **Say:** *Translate the expression to an algebraic expression.* ($84 + 2 \cdot z$, or $84 + 2z$) *In this expression, what is the variable?* (z)

Write the following on the board.

    3 times the height, minus 4

◆ **Say:** *What will the variable stand for in this expression?* (height). *Let the variable* h *stand for height. Translate this expression to an algebraic expression.* ($3h - 4$)

If students answer the Diagnostic Exercises correctly, there is probably no need for them to do this *Intervention Workshop* lesson. Have those students rejoin their class on page 475 of the student textbook. For students who have had difficulty answering the exercises above, continue the *Intervention Workshop*.

BACKGROUND

Exercises 11–16 on page 475 of the student textbook require students to translate English expressions to algebraic expressions involving variables. These *Intervention Workshop* activities address the needs of students who still have difficulty relating English expressions to numerical expressions, do not understand the concept of variables, and have difficulty writing algebraic expressions. In Activity One students write numerical expressions for English expressions. In Activity Two students write algebraic expressions for English expressions in which variables are specified. In Activity Three students identify the part of an English expression that has an unknown value, choose a variable to represent that quantity, and use the variable to write an algebraic expression. Use one or more of these activities with your students, as needed.

TEACHING THE *INTERVENTION WORKSHOP* LESSON

Getting Started

Display this figure and the text below it on the board.

Tell students that the equation is a formula for finding the perimeter of a square.

◆ **Say:** *Which parts of the formula are variables?* (*P* and *s*) *What does the variable s represent?* (the length of a side of the square) *What does the variable P represent?* (the perimeter of the square)

$$s$$

$$P = 4s$$

ACTIVITY ONE: Write a Numerical Expression for an English Expression

Write the following expression on the board.

the product of 5 and 8, minus 7

Tell students they will now use numbers and operation symbols to write a numerical expression for this English expression.

◆ **Say:** *When you find a product of two numbers, what operation do you use?* (multiplication)

◆ **Say:** *What symbols can you use to show multiplication?* ($\times$, $\cdot$) *What is another way to show multiplication?* (Write the factors next to each other, using parentheses to separate the factors.)

◆ **Say:** *What are three different ways you can write expressions to show the product of 5 and 8?* [5×8, $5 \cdot 8$, $5(8)$] *Write a numerical expression that shows the product of 5 and 8, minus 7.* ($5 \times 8 - 7$, $5 \cdot 8 - 7$, or $5(8) - 7$)

Have volunteers come to the board and write different numerical expressions for the *product of 3 and 8, minus 4.* Continue until three different ways ($3 \times 8 - 4$, $3 \cdot 8 - 4$, and $3(8) - 4$) have been listed.

Write the following expression on the board.

9 divided by 2, plus 3.

Tell students they will now write a numerical expression for this English expression.

◆ **Say:** *When you divide two numbers, what operation do you use?* (division)

◆ **Say:** *What symbol can you use to show division?* ($\div$) *What is another way to show that one number is divided by another?* (Write a fraction in which the number being divided is the numerator and the number that you are dividing by is the denominator.)

◆ **Say:** *What are two different ways you can show 9 divided by 2?* ($9 \div 2$, $9/2$) *Write a numerical expression that shows 9 divided by 2, plus 3.* ($9 \div 2 + 3$ or $9/2 + 3$)

Have volunteers write two kinds of numerical expression for 7 *divided by 3, plus 5* on the board. ($7 \div 3 + 5$, $7/3 + 5$)

Write the following expression on the board.

17 added to one half of 8

Tell students they will now write a numerical expression for this English expression.

◆ **Say:** *What operation do you use to find one half of 8?* (multiplication) *What symbol can you use to show multiplication?* (×) *What symbols can you use to show multiplication?* (×, ·) *What is another way to show multiplication?* (Write the factors next to each other using parentheses to separate the factors.)

◆ **Say:** *What are three different numerical expressions you can use to show one half of 8?* [1/2 × 8, 1/2 · 8, 1/2(8)] *What symbol do you write to show that a number is being added?* (+) *Write a numerical expression that shows 17 being added to 1/2 of 8.* (1/2 × 8 + 17, 1/2 · 8 + 17, or 1/2(8) + 17)

Have volunteers come to the board and write numerical expressions for *17 added to one half of 8.* Continue until three different ways (1/2 × 8 + 17, 1/2 · 8 + 17, and 1/2(8) + 17) have been listed.

ACTIVITY TWO: Write Algebraic Expressions

Write the following expression on the board.

n *doubled, plus 18*

Tell students they will now write an algebraic expression for this English expression. Remind students that an algebraic expression involves numbers, operation symbols, and variables. As you lead students through the translation of the expression on the board, list the algebraic equivalents of each part, as shown.

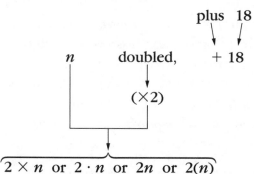

◆ **Say:** *What does it mean when you say double an amount?* (You multiply the amount by 2.)

◆ **Say:** *What symbols can you use to show multiplication?* (×, ·) *What is another way to show multiplication?* (Write the factors

next to each other.) *What are some kinds of algebraic expression you can use to show the product of 2 and the variable* n? (2 × n, 2 · n, 2n, or 2(n)) *How can you use numbers and symbols to represent plus 18.* (Write the plus sign, +, then write 18.)

◆ **Say:** *Write an algebraic expression that shows* n *doubled, plus 18.* (2 × n + 18, 2 · n + 18, 2n + 18, or 2(n) + 18)

Have volunteers write algebraic expressions for *n doubled, plus 18* on the board. Continue until the following ways have been listed:
2 × n + 18, 2 · n + 18, 2n + 18, or 2(n) + 18.

Write the following expression on the board.

z *divided by 8, minus 10*

Tell students they will now write an algebraic expression for this English expression. Remind students that an algebraic expression involves numbers, operation symbols, and variables.

◆ **Say:** *When you divide a variable by a number, what operation do you use?* (division)

◆ **Say:** *What symbol can you use to show division?* ($\div$) *What is another way to show that a variable, z, is divided by a number?* (Write a fraction in which the variable being divided is the numerator, and the number that you are dividing by is the denominator.) *What are two different ways you can show a variable, z, divided by 8?* ($z \div 8$, $z/8$)

◆ **Say:** *Write an algebraic expression that shows z divided by 8, minus 10.* ($z \div 8 - 10$ or $z/8 - 10$)

Have volunteers write an algebraic expression for *z divided by 8, minus 10* on the board. ($z \div 8 - 10$ or $z/8 - 10$)

Have students work in pairs to translate the following English expressions into algebraic expressions:

one half of *h*, added to 10
($10 + 1/2 \times h$, $10 + 1/2 \cdot h$, $10 + 1/2h$)

3 times *p*, minus 7
($3 \times p - 7$, $3 \cdot p - 7$, $3p - 7$,)

k doubled, plus 4
($2 \times k + 4$, $2 \cdot k + 4$, $2k + 4$)

5 divided by *y*, minus 9
($15 \div y - 9$, $5/y - 9$)

ACTIVITY THREE: Identify Variable Quantities and Write Algebraic Expressions

Materials: 1-10 spinner; set of 4 operations cards labeled *divided by*, *plus*, *times*, and *minus*; set of 4 variable cards with the phrases *1/2 of a number*, *the time in seconds*, *plus 6*, *a number doubled*, and the *cost in dollars*, *minus 5* for each pair of students

Write the following expression on the board.

2 times the cost in dollars, minus 10

Tell students they will now write an algebraic expression for this English expression.

◆ **Say:** *Suppose you write an algebraic expression for this expression. Will you need a variable? Why or why not?* (Yes; the cost in dollars is unknown.) *Let the variable c stand for cost. Translate the expression to an algebraic expression.* ($2c - 10$)

Write the following expression on the board.

100 minus 1/3 of the volume

Tell students they will now write an algebraic expression for this English expression.

◆ **Say:** *Suppose you write an algebraic expression for this expression. Will you need a variable? Why or why not?* (Yes; the volume is unknown.) *Choose a variable for volume. Translate the expression to an algebraic expression.* (possible response: $100 - 1/3V$)

Have students work with partners. Give each pair of students a 1-10 spinner, a set of 4 operations cards, and set of 4 variable cards. Instruct students to keep the two sets of cards separate.

Explain that students will use the spinner and index cards to write algebraic expressions. One student spins the spinner, takes an operations card and a variable card. The partner combines the number, the operations card, and variable card in any order. Partners work together to write an algebraic expression for what is shown. Partners switch roles, and repeat the activity.

SUMMING-IT-UP QUIZ

Give this quiz to students.

Write the following on the board.

> *3 times* n, *plus 5*

◆ **Say:** *Translate this expression to an algebraic expression.*
($3 \cdot n + 5$, $3n + 5$, or $3(n) + 5$)

Write the following on the board.

> *1/2 the number of tickets sold, minus 200*

◆ **Say:** *In an algebraic expression for this English expression, what would the variable stand for?*
(the number of tickets sold)

◆ **Say:** *Translate this expression to an algebraic expression.*
(Possible responses: If n is chosen as the variable, $1/2 \times n - 200$, $1/2 \cdot n - 200$, or $1/2n - 200$.)

If students answer the *Summing-It-Up Quiz* correctly, they should then resume work on page 475 of their textbook. If students do not successfully complete the *Summing-It-Up Quiz*, further remediation may be necessary.

A Manipulative Understanding of Algebra

Use with Lesson 14-4, exercises 1–4, text pages 480–481.

DIAGNOSTIC INTERVIEW

Before beginning this *Intervention Workshop* lesson, have students complete the following exercises.

Diagnostic Exercises

Write the following on the board.

$$n + 8 = 24$$

◆ **Say:** *To solve the equation, what operation do you have to do to each side of the equation?* (Subtract 8 from each side.)

Write the following on the board.

$$a - 14 = 6$$

◆ **Say:** *To solve the equation, what operation do you have to do to each side of the equation?* (Add 14 to each side.)

If students answer the Diagnostic Exercises correctly, there is probably no need for them to do this *Intervention Workshop* lesson. Have those students rejoin their class on page 481 of the student textbook. For students who have had difficulty answering the exercises above, continue the *Intervention Workshop*.

BACKGROUND

Exercises 1–4 on page 481 of the student textbook require students to choose the operation needed to solve equations involving addition and subtraction. These *Intervention Workshop* activities use manipulatives to provide an understanding of how inverse operations are chosen and used to solve equations. In Activity One students use manipulatives to represent and solve equations involving addition. In Activity Two students use manipulatives to represent and solve equations involving subtraction. In Activity Three students use manipulatives to identify the operation needed to solve equations involving addition or subtraction. Use one or more of these activities with your students, as needed.

TEACHING THE *INTERVENTION WORKSHOP* LESSON

Getting Started

Let students use connecting cubes to model the situations below.

◆ **Say:** *Suppose you start with 5 cubes, then add 2 cubes. What do you have to do next to be left with 5 cubes?* (Subtract 2 cubes.)

◆ **Say:** *Suppose you start with 6 cubes and take away 3 cubes. What do you have to do next to be left with 6 cubes?* (Add 3 cubes.)

ACTIVITY ONE: Use Manipulatives to Represent and Solve Addition Equations

Materials: 20 positive unit cards, 20 negative unit cards, and one variable card marked *n* for each pair of students; transparencies of positive unit cards, negative unit cards, and a variable card

On the overhead show a positive unit card, a negative unit card, and a variable card. Explain that a positive card represents 1, or adding 1; a negative card means subtracting 1; and the variable card stands for an unknown quantity, represented by the variable *n*.

$\boxed{+}$ ◀—— means 1, or adding 1

$\boxed{-}$ ◀—— means $^-1$, or subtracting 1

$\boxed{n}$ ◀—— the variable, *n*

On a transparency draw a box with two large sections and an equals sign connecting them. Explain that each part of the box represents one side of an equation.

In the left side of the box, place a variable card and 5 positive unit cards. In the right side of the box, place 9 positive unit cards. Then have students use their own boxes and cards to copy what you have done.

◆ **Say:** *Look at the cards on the left side of the equation. What algebraic expression do those cards represent?* (n + 5) *What do the cards on the right side of the equation represent?* (9) *What equation do the cards represent?* (n + 5 = 9)

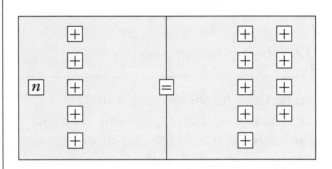

Write the equation n + *5* = *9* below the equation box.

Lead students through the process of using the manipulatives to solve the equation. Perform the steps on the overhead as students do so at their desks.

Explain that to solve the equation students must get the variable card alone on one side of the equation. Remind students that since a positive unit card and a negative card are opposites and they combine to equal zero, they cancel each other out. So to remove a positive card, you put a negative card next to it.

◆ **Say:** *Put a negative card next to each positive card on the left side of the equation box. To balance the equation, put the same number of negative cards on the right side of the equation box. Match each negative card on the right side of the box with a positive card. How many negative cards did you put on each side of the equation box?* (5) *What operation did you just perform on each side of the equation box?* (the subtraction of 5) *If a negative card is matched with a positive card, take each of the matched cards out of the box. What does the equation box show now?* ($n = 4$)

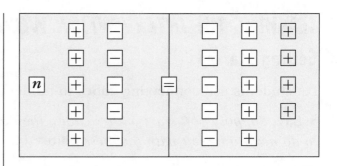

Have students work in pairs. Partners work together, using equation boxes and cards to represent each of the equations listed below. For each equation they tell what operation they need to perform to get the variable alone on one side of the equation. They then solve the equation.

$n + 8 = 19$ (subtraction of 8; $n = 11$)

$n + 6 = 10$ (subtraction of 6; $n = 4$)

$n + 3 = 12$ (subtraction of 3; $n = 9$)

ACTIVITY TWO: Use Manipulatives to Represent and Solve Addition Equations

Materials: 20 positive unit cards, 20 negative unit cards, and one variable card marked n for each pair of students; transparencies of positive unit cards, negative unit cards, and a variable card

On the overhead show a positive unit card, a negative unit card, and a variable card. Remind students that a positive card represents 1, or adding 1; a negative card means subtracting 1, and the variable card stands for an unknown quantity, represented by the variable n.

$\boxed{+}$ ⟵ means 1, or adding 1

$\boxed{-}$ ⟵ means ¯1, or subtracting 1

$\boxed{n}$ ⟵ the variable, n

Write the following on a transparency.

n − 3 = 14

Below this equation draw an equation box as shown on the following page. Then lead students through the process of using the manipulatives to represent and solve the equation. Perform the steps on the overhead as students do so at their desks.

◆ **Say:** *How can we use cards to show that* n *is on the left side of the equation?* (Put the variable card on the left side of the equation box.) *How can we show that we are subtracting 3 from* n? (Put 3 negative unit cards on the left side of the equation box.) *What should we put on the right side of the equation box?* (14 positive unit cards)

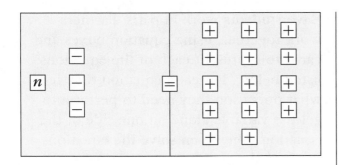

Tell students that to solve the equation they must get the variable card alone on one side of the equation. Remind students that since a positive unit card and a negative card are opposites and they combine to equal zero, they cancel each other out. So to remove a negative card, you put a positive card next to it.

◆ **Say:** *Put a positive card next to each negative card on the left side of the equation box. To balance the equation, put the same number of positive cards on the right side of the equation box. How many positive cards did you put on each side of the equation box?* (3) *What operation did you just perform on each side of the equation*

box? (the addition of 3) *If a negative card is matched with a positive card, take each of the cards out of the box. What does the equation box show now?* ($n = 17$)

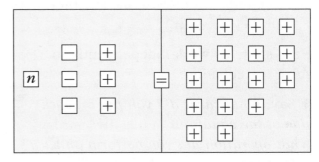

Have students work in pairs. Partners work together, using equation boxes and cards to represent each of the equations listed below. For each equation they tell what operation they need to perform to get the variable alone on one side of the equation. They then solve the equation.

$n - 7 = 5$ (addition of 7; $n = 12$)

$n - 9 = 11$ (addition of 9; $n = 20$)

$n - 4 = 12$ (addition of 4; $n = 16$)

ACTIVITY THREE: Use Manipulatives to Identify the Operation Needed to Solve an Equation

Materials: 20 positive unit cards, 20 negative unit cards, and one variable card marked n for each pair of students

Have students examine their cards. Remind students that a positive card represents 1, or adding 1; a negative card means subtracting 1, and the variable card stands for an unknown quantity, represented by the variable n.

Write the following on the board.

$n + 4 = 17$

Lead students through the process of using the manipulatives to represent the equation.

◆ **Say:** *How can we use cards to show that* n *is on the left side of the equation?* (Put the variable card on the left side of the equation box.) *How can we show that we are adding 4 to* n*?* (Put 4 positive unit cards on the left side of the equation box.) *What should we put on the right side of the equation box?* (17 positive unit cards)

Remind students that to solve the equation they must get the variable card alone on one side of the equation.

◆ **Say:** *How can you use cards to get the variable alone on one side of the equation?* (Use negative cards to cancel out positive

cards on the left side of the equation.)
When you put cards on one side of the equation, what do you have to do to balance the equation? (Put the same number and kind of card on the other side of the equation.)

Have students work independently to solve the equation.

◆ **Say:** *What cards did you put on each side of the equation?* (4 negative cards) *What operation did you perform on each side of the equation?* (the subtraction of 4) *What is* n *equal to?* (13)

Have students work in pairs. Partners work together, using equation boxes and cards to represent each of the equations listed below. For each equation they tell what operation they need to perform to get the variable alone on one side of the equation. They then solve the equation.

$n + 6 = 11$ (subtraction of 6; $n = 5$)

$n - 3 = 7$ (addition of 3; $n = 10$)

$n - 8 = 5$ (addition of 8; $n = 13$)

$n + 2 = 14$ (subtraction of 2; $n = 12$)

SUMMING-IT-UP QUIZ

Give this quiz to students.

Write the following on the board.

$n - 12 = 5$

If students wish, let them use positive and negative unit cards and an equation box to represent the equation.

◆ **Say:** *To solve the equation what do you have to do on each side of the equation?* (Add 12 to each side.)

Write the following on the board.

$a + 11 = 17$

If students wish, let them use positive and negative unit cards and an equation box to represent the equation.

◆ **Say:** *To solve the equation what do you have to do to each side of the equation?* (Subtract 11 from each side.)

If students answer the *Summing-It-Up Quiz* correctly, they should then resume work on page 481 of their textbook. If students do not successfully complete the *Summing-It-Up Quiz,* further remediation may be necessary.

Use with Lesson 14-12, exercises 1–16, text pages 496–497.

DIAGNOSTIC INTERVIEW

Before beginning this *Intervention Workshop* lesson, have students complete the following exercises.

Diagnostic Exercises

◆ **Say:** *Write these pairs of numbers. Then compare the numbers by writing <, =, or >.*

> $1/5$ *and* 0.2 0.34 *and* $2/3$ $3/5$ *and* 3.5

$(1/5 = 0.2; 0.34 < 2/3; 3/5 < 3.5)$

◆ **Say:** *Write these numbers in order, from least to greatest.*

> $1/2, 8/10, 0.23,$ *and* 0.9

$(0.23, 1/2, 8/10, 0.9)$

If students answer the Diagnostic Exercises correctly, there is probably no need for them to do this *Intervention Workshop* lesson. Have those students rejoin their class on page 497 of the student textbook. For students who have had difficulty answering the exercises above, continue the *Intervention Workshop*.

BACKGROUND

This *Intervention Workshop* lesson addresses the problem of understanding that a number line can show all the positive numbers: whole numbers, fractions, and decimals. Students begin with using patterns to label missing fractions and decimal tick marks on a number line. Then they compare fractions, decimals, and whole numbers on a number line. The scales on the number lines are labeled in such a way as to guide students to distinguish fractions from decimals and help them compare the numbers. Finally students order fractions and decimals on a number line.

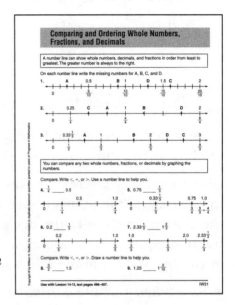

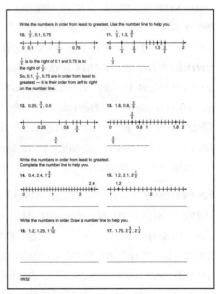

TEACHING THE *INTERVENTION WORKSHOP* LESSON

Getting Started

Materials: number lines (BLM 19)

◆ **Say:** *Write the following. Then write the next three numbers in each pattern.*

    0.1, 0.2, 0.3, 0.4   1/9, 2/9, 3/9, 4/9

(0.5, 0.6, 0.7; 5/9, 6/9, 7/9)

◆ **Say:** *Write these numbers. Write the fraction or decimal equivalent for each.*

    0.5   2/3   8/2   2.25

(1/2; 0.66 2/3; 4.0; 9/4)

◐ USING PAGES IW31 and IW32

Have a student read aloud the material in the box at the top of page IW31.

Direct students' attention to the labeled tick marks on the number lines. Relate the statement given to the decimal and fraction equivalents students wrote earlier in *Getting Started* in the following way.

◆ **Say:** *When you use a number line to compare numbers it is useful to write both the fraction and decimal equivalents for each of the tick marks shown. On the number lines in this lesson the fractions are written below the tick marks and the decimals are written above the tick marks. This will help you distinguish fractions from decimals and help you see which number is greater.*

Exercises 1–9 on page IW31

Have a student read aloud the material in the box in the middle of the page. Then have the student read aloud the direction line above exercises 1–3. Assist students as necessary to complete these exercises.

Have a student read aloud the direction line above exercises 4–7. Assist students as necessary to interpret the graphs of the numbers as shown.

◆ **Say:** *Remember that the arrow symbols for* is less than *and* is greater than *always point to the lesser number.*

Assist students who have difficulty with exercise 5.

◆ **Say:** *Remember that 3/3 equals 1 and 4/4 equals 1, so 3/3 equals 4/4.*

Have a student read aloud the direction line above exercises 8–9. Guide students as necessary to draw their own number lines to compare the numbers given.

By analyzing the graphs and making their own graphs of numbers on a number line, exercises 1–9 guide students to compare whole numbers, fractions, and decimals.

Exercises 10–17 on page IW32
Have a student read aloud the direction line at the top of page IW32.

◆ **Say:** *You can easily see how to write numbers in order by looking at their order on a number line. Remember that the greater number is always to the right.*

Guide students through the solution given for exercise 10. Assist students as necessary to complete exercises 11–13.

You can point out how the fractions and decimals are written to label the tick marks given. Then have a student read aloud the direction line above exercises 14–15. Assist students as necessary to graph the remaining numbers on the number lines given.

Finally have a student read aloud the direction line above exercises 16–17. Guide students to draw their own number lines to write the numbers in order, from least to greatest.

Exercises 10–17 guide students to order numbers from least to greatest on a number line.

SUMMING-IT-UP QUIZ

Give this quiz to students.

◆ **Say:** *Write these pairs of numbers. Compare the numbers by writing < =, or >.*

        5/4 and 1.25   0.33 1/3 and 2/3   7/5 and 0.75

($5/4 = 1.25$; $0.33 \; 1/3 < 2/3$; $7/5 > 0.75$)

◆ **Say:** *Write these numbers in order, from least to greatest.*

        3 1/2, 3.2, 3/2, 0.32

($0.32, 3/2, 3.2, 3 \; 1/2$)

If students answer the *Summing-It-Up Quiz* correctly, then they should return to Lesson 14-12 and resume work on page 497 of their textbook. If students do not successfully complete the *Summing-It-Up Quiz,* further remediation may be necessary.

Answers: Pages IW31 and IW32

1. A: 2/10 = 0.2; B: 9/10 = 0.9; C: 16/10 = 1.6; D: 13/10 = 1.3

2. A: 3/4 5 0.75; B: 5/4 5 1.25; C: 2/4 5 0.5; D: 7/4 5 1.75

3. A: 2/3 = 0.66 2/3; B: 5/3 = 1.66 2/3; C: 8/3 = 2.66 2/3; D: 7/3 = 2.33 1/3

4. <	**5.** >	**6.** =	**7.** >	**8.** =	**9.** >

11. 1.3, 3/2   **12.** 0.25, 0.6   **13.** 0.8, 1.8

14. 0.4, 1 3/4, 2.4   **15.** 1.2, 2.1, 2 1/2   **16.** 1.2, 1.25, 1 9/10

17. 1.75, 2 1/4, 2 3/4

Section II

Teacher's Edition
Intervention Workshop

Student Pages

Understanding Place Value

> **Digits** are the building blocks of numbers. With these digits
>
> 0   1   2   3   4   5   6   7   8   9
>
> you can write any number.

How are these numbers the *same* and how are they *different*?

1. 54 and 45

Same: The numbers each have ___2___ digits.

The numbers contain the same digits: ___4___ and _____.

Different: The digits in the numbers are reversed.

$$54 = \underline{\quad 5 \text{ tens} \quad} + \underline{\quad 4 \text{ ones} \quad}$$
$$45 = \underline{\qquad\qquad} + \underline{\qquad\qquad}$$

2. 540, 450, 405, and 504

Same: The numbers each have _____ digits.

The numbers contain the same digits: _____, _____ and _____.

Different: The digits in the numbers are in different places.

$$540 = \underline{\;5 \text{ hundreds}\;} + \underline{\;4 \text{ tens}\;} + \underline{\;0 \text{ ones}\;}$$
$$450 = \underline{\qquad\quad} + \underline{\quad\;\;} + \underline{\quad\;\;}$$
$$405 = \underline{\qquad\quad} + \underline{\quad\;\;} + \underline{\quad\;\;}$$
$$504 = \underline{\qquad\quad} + \underline{\quad\;\;} + \underline{\quad\;\;}$$

3. 54, 504, and 5004

Same: The numbers start with the same digit: _____.

The digit ___4___ is always in the same place, the _____ place.

Different: The numbers have a different number of places. The digit 5 is in different places.

In 54, 5 is in the ___tens___ place.

In 504, 5 is in the _____ place.

In 5004, 5 is in the _____ place.

4. 40,500; 40,050; 40,005; and 45,000

Same: _____

_____

_____

Different: _____

_____

_____

A **place-value chart** can help you understand the places and values of each digit in a number, especially a large number.

Period →	Billions			Millions			Thousands			Ones		
Place →	Hundred Billions	Ten Billions	One Billions	Hundred Millions	Ten Millions	One Millions	Hundred Thousands	Ten Thousands	One Thousands	Hundreds	Tens	Ones

Name the place and value of the digit 3 in each number.

5. 370

Hundreds	Tens	Ones
3	7	0

Place: ___hundreds___

Value: 3×100 or _____

6. 703

Hundreds	Tens	Ones
7	0	3

Place: _____

Value: 3×1 or _____

7. 7003

Thousands	Hundreds	Tens	Ones
7	0	0	3

Place: _____

Value: $3 \times$ _____ or _____

8. 3700

Thousands	Hundreds	Tens	Ones
3	7	0	0

Place: _____

Value: 3×1000 or _____

9. 70,030

Ten Thousands	Thousands	Hundreds	Tens	Ones
7	0	0	3	0

Place: _____

Value: _____

10. 37,000

Ten Thousands	Thousands	Hundreds	Tens	Ones
3	7	0	0	0

Place: _____

Value: _____

11. 73,000

Ten Thousands	Thousands	Hundreds	Tens	Ones
7	3	0	0	0

Place: _____

Value: _____

12. 730,000

Hundred Thousands	Ten Thousands	Thousands	Hundreds	Tens	Ones

Place: _____   Value: _____

13. 300,070

Hundred Thousands	Ten Thousands	Thousands	Hundreds	Tens	Ones

Place: _____   Value: _____

Estimating with Compatible Numbers

> If it is easy to divide one number by another mentally, the numbers are **compatible numbers**.

Which pairs of numbers are compatible? If the numbers are **not** compatible, find the nearest pair of compatible numbers.

1. 36 and 9

You *can* separate 36 into equal groups of 9 with none left over.

The number of equal groups is _____.
36 and 9 are compatible because 36 can be divided by 9.

$36 \div 9 =$ _____.

You can use patterns to make a family of compatible numbers.

36 and 9	$36 \div 9 =$ _____
3**60** and 9	$360 \div 9 =$ __**40**__
3**600** and 9	$3600 \div 90 =$ _____
3**6,000** and 9	$36{,}000 \div 90 =$ _____

---

2. 55 and 8

You *cannot* separate 55 into equal groups of 8 with none left over.

← You need 1 more.

Compatible numbers near 55 and 8

are ___8___ and _____.

Use patterns to make a family of compatible numbers.

_____ and 8	_____ $\div$ 8 = _____
_____**0** and 8	_____**0** $\div$ 8 = _____
_____**00** and **80**	_____**00** $\div$ **80** = _____
_____**,000** and **80**	_____**000** $\div$ **80** = _____

3. 46 and 6

You _____ separate 46 into equal groups of 6 with none left over. Complete the diagram to answer.

☐ ☐ ☐ ☐ ☐ ☐

Use patterns to make a family of compatible numbers.

_____ and 6   _____ ÷ 6 = _____

_____0 and 6   _____0 ÷ 6 = _____

_____00 and **60**   _____00 ÷ 60 = _____

_____,000 and **60**   _____000 ÷ 60 = _____

Compatible numbers near 46 and 6

are _____ and ___6___ .

Use compatible numbers to estimate the quotient.

4. 390 ÷ 8

Nearby compatible numbers for 39 and 8

are _____ and 8 because

_____ ÷ 8 = _____.

Complete the pattern to find the estimated quotient.

_____ ÷ 8 = _____

_____0 ÷ 8 = _____**0**

5. 842 ÷ 9

Nearby compatible numbers for 84 and 9

are _____ and 9 because

_____ ÷ 9 = _____.

Complete the pattern to find the estimated quotient.

_____ ÷ 9 = _____

_____0 ÷ 9 = _____**0**

6. 2784 ÷ 52

Nearby compatible numbers for 278 and 52

are _____ and 50 because

_____ ÷ 50 = _____.

Complete the pattern to find the estimated quotient.

_____ ÷ 50 = _____

_____0 ÷ 50 = _____**0**

7. 7686 ÷ 39

Nearby compatible numbers for 769 and 39

are _____ and 40 because

_____ ÷ 40 = _____.

Complete the pattern to find the estimated quotient.

_____ ÷ 40 = _____

_____0 ÷ 40 = _____**0**

Understanding Decimal Place Value

You can name fractions with denominators of 10, 100, 1000 etc., as **decimals**.

This place-value chart is extended to the right of the ones place to show **decimal place value**.

Place →

Tens	Ones	Tenths	Hundredths	Thousandths
	0 ●	$\frac{1}{10} = 0.1$	$\frac{1}{100} = 0.01$	$\frac{1}{1000} = 0.001$

↑ decimal point

The *numerator* of the fraction is written to the right of the decimal point.

The *denominator* of the fraction is expressed by the place of the digits in the decimals.

Write each fraction as a decimal. Then complete.

1. $\frac{3}{10}$

Ones	Tenths
0 ●	3

_____ = ____0.3____

2. $\frac{3}{100}$

Ones	Tenths	Hundredths
0 ●	0	_____

You need a 0 as a placeholder.

_____ = ____0.0____

3. $\frac{3}{1000}$

Ones	Tenths	Hundredths	Thousandths
0 ●			

You need _____ zeros as placeholders.

_____ = ____0.____

4. $\frac{35}{1000}$

Ones	Tenths	Hundredths	Thousandths
0 ●			

You need _____ zero as a placeholder.

_____ = _____

5. $4\frac{35}{100}$

Tens	Ones	Tenths	Hundredths	Thousandths
	●			

You need _____ zeros as placeholders.

_____ = _____

6. $15\frac{5}{1000}$

Tens	Ones	Tenths	Hundredths	Thousandths
	●			

You need _____ zeros as placeholders.

_____ = _____

This place-value chart shows additional **decimal place-values**.

Ones	Tenths	Hundredths	Thousandths	Ten-Thousandths	Hundred-Thousandths
0	$\frac{1}{10} = 0.1$	$\frac{1}{100} = 0.01$	$\frac{1}{1000} = 0.001$	$\frac{1}{10,000} = 0.0001$	$\frac{1}{100,000} = 0.00001$

↑ decimal point

Name the place value and the value of the digit **7** in each number.

7. 0.73

Ones	Tenths	Hundredths
0	7	3

Place: ___Tenths___

Value: 7 × 0.1 or _____

8. 7.03

Ones	Tenths	Hundredths
7	0	3

Place: _____

Value: 7 × 1 or _____

9. 0.007

Ones	Tenths	Hundredths	Thousandths
0	0	0	7

Place: _____

Value: 7 × 0.001 or _____

10. 3.07

Ones	Tenths	Hundredths
3	0	7

Place: _____

Value: 7 × 0.01 or _____

11. 0.0307

Ones	Tenths	Hundredths	Thousandths	Ten-Thousandths

Place: _____

Value: _____

12. 30.07

Tens	Ones	Tenths	Hundredths

Place: _____

Value: _____

13. 0.307

Ones	Tenths	Hundredths	Thousandths

Place: _____

Value: _____

14. 0.00307

Ones	Tenths	Hundredths	Thousandths	Ten-Thousandths	Hundred-Thousandths

Place: _____   Value: _____

15. 3.03703

Ones	Tenths	Hundredths	Thousandths	Ten-Thousandths	Hundred-Thousandths

Place: _____   Value: _____

Comparing and Ordering Decimals

If you divide into 10 equal parts, each part is 1 tenth.

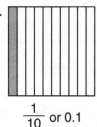

$\frac{1}{10}$ or 0.1

If you divide 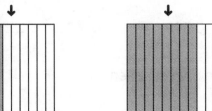 into 10 equal parts, there are 100 equal parts. Each part is 1 hundredth.

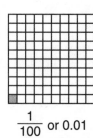

$\frac{1}{100}$ or 0.01

Use the models to compare. Write <, =, or >.
(*Remember:* The arrow in the symbols for *is greater than* and *is less than* always points to the lesser number.)

Think: 0.6 is the same as 0.60.

1.   0.4 _____<_____ 0.8
          ↓   ↓

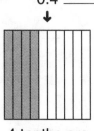

4 tenths are shaded.   8 tenths are shaded.

2.   0.65 _____ 0.6
          ↓   ↓

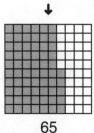

65 hundredths are shaded.   6 tenths are shaded.

Draw a model for the second number. Then compare, using <, =, or >.

3.   1.2 _____ 1.20
          ↓   ↓

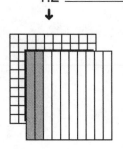

4.   4.3 _____ 4.03
          ↓   ↓

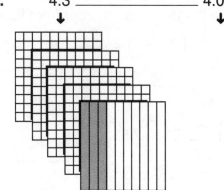

A **number line** shows numbers in order from least to greatest, from left to right.
On this number line each decimal shown is 1 tenth greater than the decimal to its left.

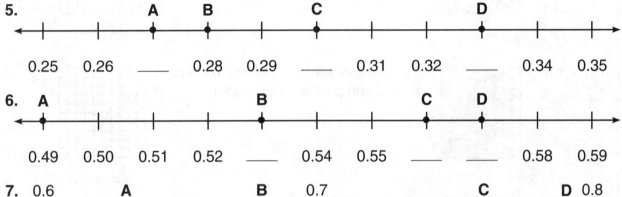

0.0   0.1   0.2   0.3   0.4   0.5   0.6   0.7   0.8   0.9   1.0   1.1

Write the missing numbers for A, B, C, and D on each number line.

5.

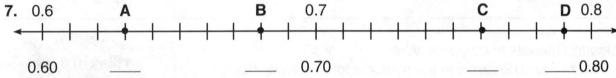

0.25   0.26   ____   0.28   0.29   ____   0.31   0.32   ____   0.34   0.35

6.

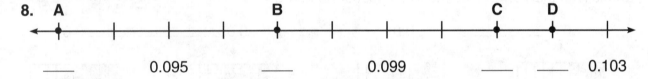

0.49   0.50   0.51   0.52   ____   0.54   0.55   ____   ____   0.58   0.59

7. 0.6   A   B   0.7   C   D 0.8

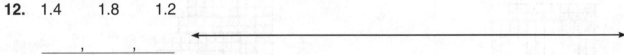

0.60   ____   ____   0.70   ____   ____ 0.80

8.   A   B   C   D

____   0.095   ____   0.099   ____   ____   0.103

Study each pattern. Then continue the pattern in the spaces given.

9.   4.1   4.2   4.3   4.4   _____   __4.6__   _____

10.   5.65   5.66   5.67   5.68   _____   _____   _____

11.   2.180   2.181   2.182   2.183   _____   _____   _____

Mark a number line and place the decimals on it.
Then write the decimals in order from least to greatest.

12.   1.4   1.8   1.2

_____ , _____ , _____

13.   2.51   2.61   2.52

_____ , _____ , _____

Multiplication: Placing the Decimal Point

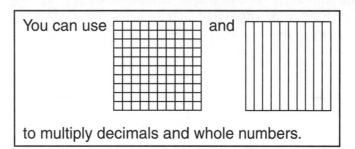
You can use [grid] and [grid] to multiply decimals and whole numbers.

Multiply.

1. 3×0.4 *Remember:* 3 times 0.4 is the same as $0.4 + 0.4 + 0.4$.

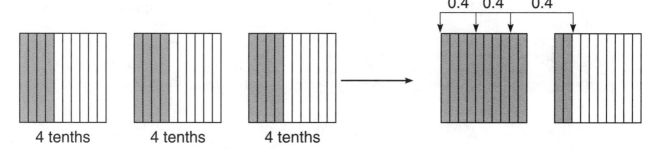

4 tenths   4 tenths   4 tenths

$3 \times 0.4 =$ _____ whole + _____ tenths, or _____

Complete the models to help you multiply.

2. 2×1.6

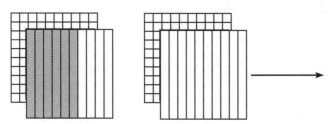

$2 \times 1.6 =$ _____ wholes + _____ tenths, or _____

3. 4×0.35

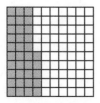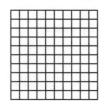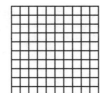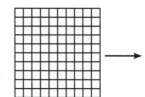

35 hundredths

$4 \times 0.35 =$ _____

You can multiply two decimals by modeling with hundred flats. Let the columns of the flats stand for one decimal, and the rows stand for the other decimal. (*Hint:* columns are vertical; they go up and down. Rows are horizontal; they go across.)

10 Rows →

↑ 10 Columns

Multiply.

4. 0.4×2.3

Show 2.3 with columns.

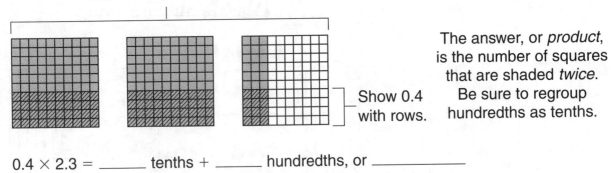

Show 0.4 with rows.

The answer, or *product*, is the number of squares that are shaded *twice*. Be sure to regroup hundredths as tenths.

$0.4 \times 2.3 = $ _____ tenths + _____ hundredths, or _____

The columns for 3.6 are shaded in. Shade in rows for 0.6. Then use your model to find the product.

5. 0.6×3.6

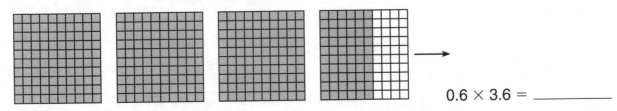

$0.6 \times 3.6 = $ _____

To find out how many decimal places a decimal has, count the number of digits to the *right* of the decimal point.

18.0506

4 decimal places

5.600

3 decimal places

Tell how many decimal places.

6. 0.52

2 decimal places

7. 10.005

_____

8. 18.6060

_____

Division: Placing the Decimal Point

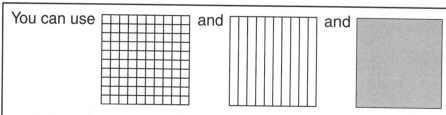

You can use [grid] and [rods] and [shaded square] to divide decimals by whole numbers and decimals by decimals.

Divide.

1. 2.1 ÷ 3

Divide ___2.1___ into ___3___ equal groups.

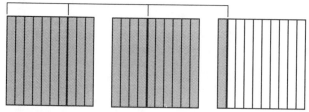

Check: 0.7 + 0.7 + 0.7 = 2.1,

so 2.1 ÷ 3 = ___0.7___

2. 2.4 ÷ 4

Divide _____ into _____ equal groups.

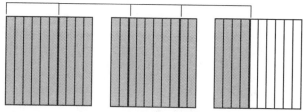

Check: 0.6 + 0.6 + 0.6 + 0.6 = 2.4,

so 2.4 ÷ 4 = _____

In exercises 3–5 complete the models to help you divide.

3. 1.25 ÷ 5

Divide _____ into _____ equal groups.

 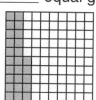

1.25 ÷ 5 = _____

4. 0.28 ÷ 2

Divide _____ into _____ equal groups.

0.28 ÷ 2 = _____

5. 3.15 ÷ 3

Divide _____ into _____ equal groups.

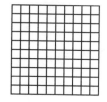 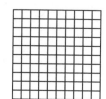 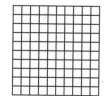

3.15 ÷ 3 = _____

Divide using the model. Then complete the pattern. Notice that the *quotients*, or answers, in each pattern are the same.

6. 1.5 ÷ 0.5

Divide __1.5__ into equal groups of __0.5__ .

1.5 ÷ 0.5 = _____3_____

15 ÷ 5 = _____

150 ÷ 50 = _____

7. 3 ÷ 0.6

Divide _____ into equal groups of _____ .

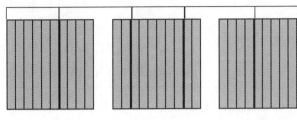

3 ÷ 0.6 = _____

30 ÷ 6 = _____

300 ÷ 60 = _____

8. 0.36 ÷ 0.06

Divide _____ into equal groups of _____ .

0.36 ÷ 0.06 = _____

3.6 ÷ 0.6 = _____

36 ÷ 6 = _____

9. 3.6 ÷ 0.6

Divide _____ into equal groups of _____ .

3.6 ÷ 0.6 = _____

36 ÷ 6 = _____

360 ÷ 60 = _____

Complete the pattern.

10. 8400 ÷ 200 = __42__

840 ÷ 20 = __42__

84 ÷ 2 = _____

8.4 ÷ 0.2 = _____

0.84 ÷ 0.02 = _____

11. 61,800 ÷ 300 = _____

6180 ÷ 30 = _____

618 ÷ 3 = _____

61.8 ÷ 0.3 = _____

6.18 ÷ 0.03 = _____

12. 100,050 ÷ 50 = _____

10,005 ÷ 5 = _____

1000.5 ÷ 0.5 = _____

100.05 ÷ 0.05 = _____

10.005 ÷ 0.005 = _____

Understanding Factoring

If you can make *only one* rectangular array for a number, the number is a **prime number.**

If you can make *more than one* rectangular array for a number, the number is a **composite number.**

Remember: Changing the position of an array does not make a different array: □ □ □ □ is the same as □ □
□ □ □ □   □ □
                                                                                                  □ □
                                                                                                  □ □

Tell whether the number is prime or composite.

1. 12

□ □ □ □ □ □ □ □ □ □ □ □   □ □ □ □ □   □ □ □ □
                           □ □ □ □ □   □ □ □ □
                                         □ □ □ □

___12___ is a ___composite___ number.

2. 13

□ □ □ □ □ □ □ □ □ □ □ □ □

You can make only _____ array for _____13_____.

___13___ is a _____ number.

Complete the array(s) to help you decide whether the number is prime or composite. Draw all possible arrays for each.

3. 24

□ □

_____ is a _____ number.

4. 32

□ □

_____ is a _____ number.

5. Find the prime numbers to 100 by following these steps.

1	2	3	4	5	6	7	8	9	10
11	12	13	14	15	16	17	18	19	20
21	22	23	24	25	26	27	28	29	30
31	32	33	34	35	36	37	38	39	40
41	42	43	44	45	46	47	48	49	50
51	52	53	54	55	56	57	58	59	60
61	62	63	64	65	66	67	68	69	70
71	72	73	74	75	76	77	78	79	80
81	82	83	84	85	86	87	88	89	90
91	92	93	94	95	96	97	98	99	100

Step 1: Cross out 1. 1 is not prime or composite.
Step 2: 2 is prime. Ring 2. Cross out every second number after 2.
Step 3: 3 is prime. Ring 3. Cross out every third number after 3.
Step 4: 5 is prime. Ring 5. Cross out every fifth number after 5.
Step 5: 7 is prime. Ring 7. Cross out every seventh number after 7.

The numbers that are ringed or not crossed out are prime numbers.
List the prime numbers to 100:

2, 3, 5, 7, _____

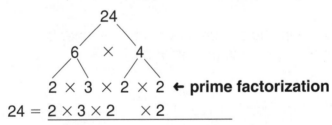

You can find the **prime factors** of a composite number by using a **factor tree**.

24
6 × 4
2 × 3 × 2 × 2 ← **prime factorization**
24 = 2 × 3 × 2 __× 2__

Complete the factor tree to help you find the **prime factorization**.

6.

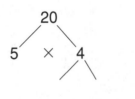

20
5 × 4

_____ × _____

20 = _____ × _____ × _____

7.

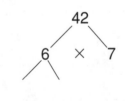

42
6 × 7

_____ × _____

42 = _____ × _____ × _____

Greatest Common Factor

You can draw rectangles on grid paper to find the possible factors of a number.
Factors are numbers you multiply together to give another number, called the **product**.

Remember: is the same as

Use the rectangles. Find all the factors of the numbers.

1. 12

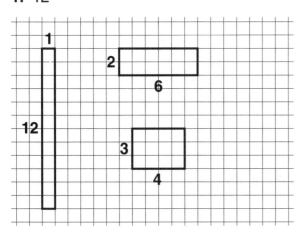

2. 10

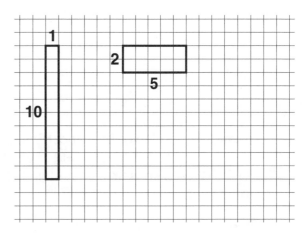

Factors of 12: 1, 2, 3, _____

Factors of 10: 1, _____ ,10

Complete the drawings of rectangles to help you find all the factors.

3. 16

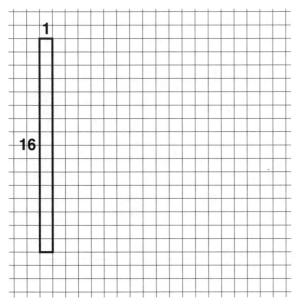

4. 18

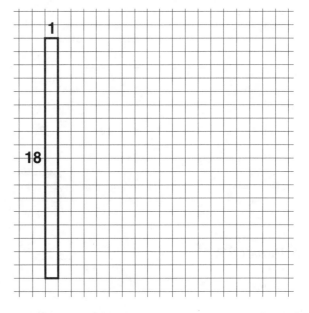

Factors of 16: _____

Factors of 18: _____

You can find the **common factors** of two numbers by drawing rectangles.

Draw all the possible rectangles for each number.
Then compare the sets of factors to find the common factors.

5. 9 and 15

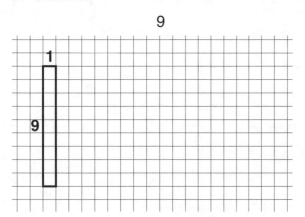

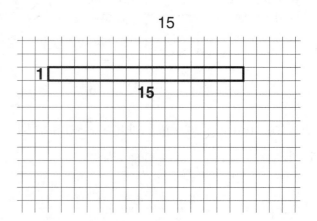

Factors of 9: _____   Factors of 15: _____

Common factors of 9 and 15: 1,_____

The greatest common factor of 9 and 15 is: _____.

Find the GCF. Draw all possible rectangles for both numbers
in order to answer.

6. 21 and 24

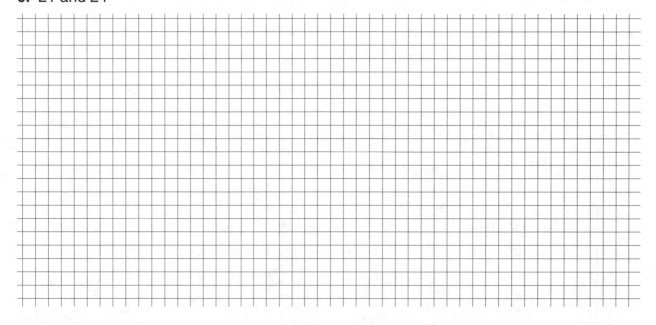

Factors of 21: _____   Factors of 24: _____

Common factors of 21 and 24: _____

The greatest common factor of 21 and 24 is: _____.

Finding Equivalent Fractions

Fractions that name the same part of a set, region, or object are called **equivalent fractions**. You can use grid paper to find *equivalent fractions*.

Complete the number sentence that shows equivalent fractions.

1. The shaded areas are the same, so the fractions are equivalent.

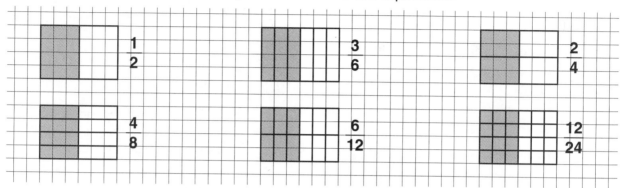

$$\frac{1}{2} = \frac{3}{6} = \frac{2}{4} = \frac{4}{8} = \frac{6}{12} = \frac{12}{24}$$

2. The shaded areas are the same, so the fractions are equivalent.

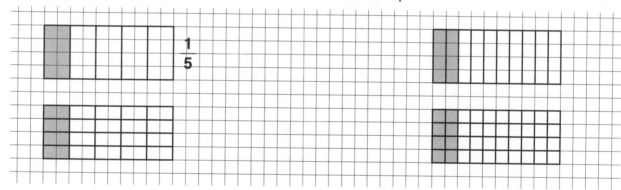

$$\frac{1}{5} = \frac{2}{?} = \frac{?}{?} = \frac{?}{?}$$

Complete the pattern on the fraction tape to find equivalent fractions.

3.

_____	$\frac{4}{6}$	$\frac{6}{9}$	$\frac{8}{12}$	$\frac{10}{15}$	_____	_____	_____

For each fraction, make a fraction tape. Tell the fraction that is at the left end of the tape.

4. $\frac{4}{16}$ **5.** $\frac{5}{15}$ **6.** $\frac{10}{16}$

You can use fraction strips to add and subtract fractions.

Add or subtract. Use the fraction strips to help you. Simplify where necessary.

7. $\frac{1}{3} + \frac{1}{3}$

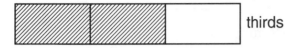 thirds

$\frac{1}{3} + \frac{1}{3} =$ _____

8. $\frac{3}{8} + \frac{5}{8}$

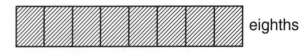 eighths

$\frac{3}{8} + \frac{5}{8} = \frac{8}{8} =$ _____

Add or subtract. Complete the fraction strips to help you.

9. $\frac{3}{4} - \frac{1}{4}$

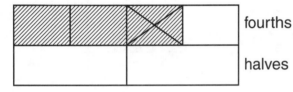 fourths

halves

$\frac{3}{4} - \frac{1}{4} = \frac{2}{4} =$ _____

10. $\frac{7}{10} - \frac{3}{10}$

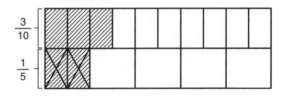 tenths

fifths

$\frac{7}{10} - \frac{3}{10} =$ _____ $=$ _____

You cannot add or subtract fractions with unlike denominators. To add or subtract these fractions, you need to rename them with the same, or **common**, **denominator**.

Add or subtract. Use the fraction strips to help you.

11. $\frac{1}{4} + \frac{1}{2}$

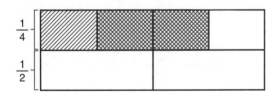

$\frac{1}{4} + \frac{1}{2} = \frac{1}{4} + \frac{2}{4} =$ _____

12. $\frac{3}{10} - \frac{1}{5}$

$\frac{3}{10} - \frac{1}{5} = \frac{3}{10} -$ _____ $=$ _____

Complete the fraction strips to help you add or subtract.

13. $\frac{3}{4}$

$\frac{1}{6}$

$\frac{3}{4} + \frac{1}{6} =$ _____ $=$ _____

14. $\frac{2}{3}$

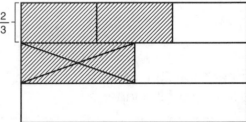

$\frac{2}{3} - \frac{1}{2} =$ _____ $=$ _____

Meaning of Mixed Numbers

A number that has a whole number part and a fraction part is a **mixed number**. For example, $2\frac{3}{4}$ is a mixed number. You can use grid paper to show mixed numbers.

Name the mixed number shown. Simplify the fraction part if you can.

1.

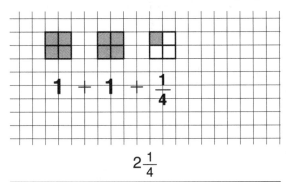

$$2\frac{1}{4}$$

2.

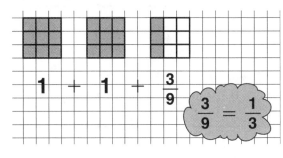

3.

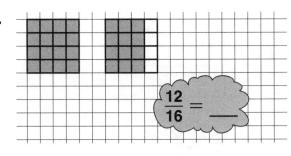

4.

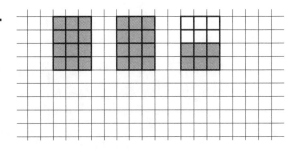

5. Match each model with the mixed number that the model shows.

a)

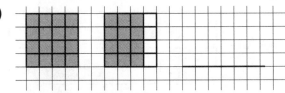

$$1\frac{3}{4}$$

b)

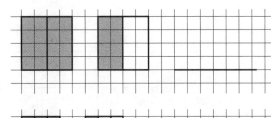

c)

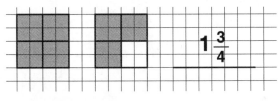

d)

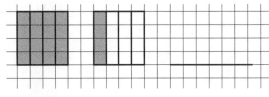

e)

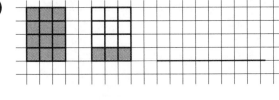

f)

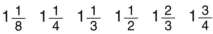

$$1\frac{1}{8} \quad 1\frac{1}{4} \quad 1\frac{1}{3} \quad 1\frac{1}{2} \quad 1\frac{2}{3} \quad 1\frac{3}{4}$$

Use with Lesson 6-4, text pages 208–209.

You can use a number line to show mixed numbers and **improper fractions**.
Remember: An improper fraction is a fraction with a numerator equal to or greater than the denominator. For example, $\frac{8}{8}$ and $\frac{13}{8}$ are improper fractions.

Write the missing mixed number and improper fraction for A and B.

6.

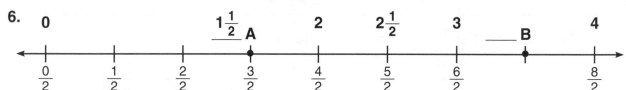

7.

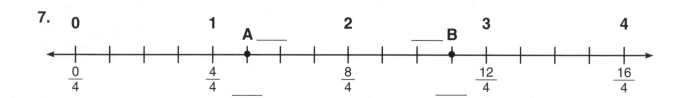

Write the mixed number shown by the fraction strips in simplest form.

8.

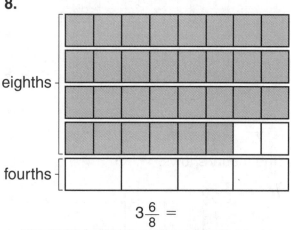

eighths

fourths

$3\frac{6}{8} =$

_____

9.

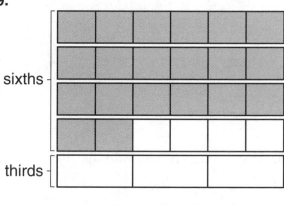

sixths

thirds

_____

10.

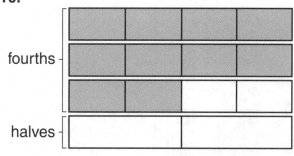

fourths

halves

_____

11.

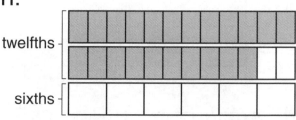

twelfths

sixths

_____

Multiplying Fractions

You can use rectangles to multiply fractions. Show one fraction along the length of the rectangle, and the other fraction along the width. (*Remember:* $\frac{1}{4}$ of $\frac{1}{2}$ means $\frac{1}{4} \times \frac{1}{2}$.)

Multiply.

1. $\frac{1}{4} \times \frac{3}{5} = \frac{3}{20}$

Divide the length into *fifths* and show $\frac{3}{5}$.

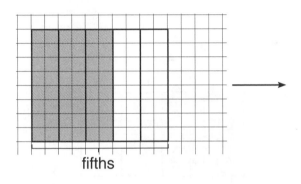

fifths

Divide the width into *fourths* and show $\frac{1}{4}$ of $\frac{3}{5}$.

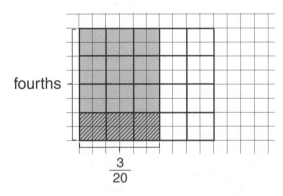

fourths

$\frac{3}{20}$

2. $\frac{1}{2} \times \frac{3}{4} =$ _____

Divide the length into *fourths* and show $\frac{3}{4}$.

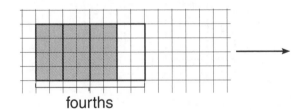

fourths

Divide the width into halves and show $\frac{1}{2}$ of $\frac{3}{4}$.

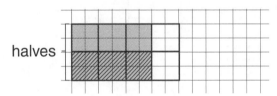

halves

Complete the model to help you multiply.

3. $\frac{1}{2} \times \frac{5}{6} =$ _____

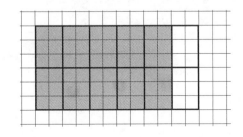

4. $\frac{2}{3} \times \frac{2}{5} =$ _____

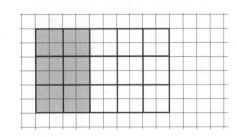

Complete the pattern to find the product.

5. $\frac{1}{2} \times \frac{1}{2} = \frac{1}{4}$ $\frac{1}{2} \times \frac{1}{6} = $ _____

  $\frac{1}{2} \times \frac{1}{3} = \frac{1}{6}$ $\frac{1}{2} \times \frac{1}{7} = $ _____

  $\frac{1}{2} \times \frac{1}{4} = \frac{1}{8}$ $\frac{1}{2} \times \frac{1}{8} = $ _____

  $\frac{1}{2} \times \frac{1}{5} = \frac{1}{10}$ $\frac{1}{2} \times \frac{1}{9} = $ _____

6. $\frac{1}{3} \times \frac{1}{2} = \frac{1}{6}$ $\frac{1}{3} \times \frac{1}{6} = $ _____

  $\frac{1}{3} \times \frac{1}{3} = \frac{1}{9}$ $\frac{1}{3} \times \frac{1}{7} = $ _____

  $\frac{1}{3} \times \frac{1}{4} = \frac{1}{12}$ $\frac{1}{3} \times \frac{1}{8} = $ _____

  $\frac{1}{3} \times \frac{1}{5} = \frac{1}{15}$ $\frac{1}{3} \times \frac{1}{9} = $ _____

Complete the model to find the product.

7. $\frac{1}{2} \times \frac{2}{3} = $ _____

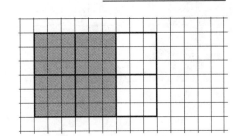

8. $\frac{1}{3} \times \frac{3}{4} = $ _____

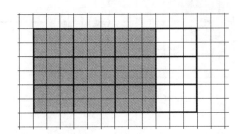

9. $\frac{2}{3} \times \frac{3}{5} = $ _____

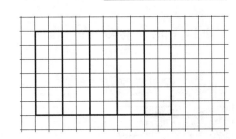

10. $\frac{1}{6} \times \frac{3}{4} = $ _____

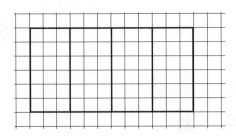

Complete the pattern to find the product.

11. $\frac{1}{2} \times \frac{2}{3} = \frac{1}{3}$ $\frac{1}{2} \times \frac{2}{7} = $ _____

  $\frac{1}{2} \times \frac{2}{4} = \frac{1}{4}$ $\frac{1}{2} \times \frac{2}{8} = $ _____

  $\frac{1}{2} \times \frac{2}{5} = \frac{1}{5}$ $\frac{1}{2} \times \frac{2}{9} = $ _____

  $\frac{1}{2} \times \frac{2}{6} = \frac{1}{6}$

12. $\frac{1}{3} \times \frac{3}{4} = \frac{1}{4}$ $\frac{1}{3} \times \frac{3}{7} = $ _____

  $\frac{1}{3} \times \frac{3}{5} = \frac{1}{5}$ $\frac{1}{3} \times \frac{3}{8} = $ _____

  $\frac{1}{3} \times \frac{3}{6} = \frac{1}{6}$ $\frac{1}{3} \times \frac{3}{9} = $ _____

Mean and Median

The **mean** of a set of data is the sum of the data items divided by the number of items.

You can use models to find the mean of different numbers. Regroup the sum of the numbers into equal groups. The **mean** is the number in each group.

1. 5   8   4   7

5: ☐ ☐ ☐ ☐ ☐   ⟶   ☐ ☐ ☐ ☐ ☐ ☐

8: ☐ ☐ ☐ ☐ ☐ ☐ ☐ ☐   ⟶   ☐ ☐ ☐ ☐ ☐ ☐

4: ☐ ☐ ☐ ☐   ⟶   ☐ ☐ ☐ ☐ ☐ ☐

7: ☐ ☐ ☐ ☐ ☐ ☐ ☐   ⟶   ☐ ☐ ☐ ☐ ☐ ☐

⬆ Four unequal groups.   ⬆ Four equal groups with the same sum.

The mean of 5, 8, 4, and 7 is ___6___.

Find the mean. Complete the model to help you.

2. 3   7   6   4   5

3: ☐ ☐ ☐   ⟶   ☐

7: ☐ ☐ ☐ ☐ ☐ ☐ ☐   ⟶   ☐

6: ☐ ☐ ☐ ☐ ☐ ☐   ⟶   ☐

4: ☐ ☐ ☐ ☐   ⟶   ☐

5: ☐ ☐ ☐ ☐ ☐   ⟶   ☐

Mean: _____

3. 9   12   13   12   14   14   13   9   9   15

☐ ☐ ☐ ☐ ☐ ☐ ☐ ☐

☐ ☐ ☐ ☐ ☐ ☐ ☐ ☐ ☐ ☐

☐ ☐ ☐ ☐ ☐ ☐ ☐ ☐ ☐ ☐ ☐

☐ ☐ ☐ ☐ ☐ ☐ ☐ ☐ ☐ ☐

☐ ☐ ☐ ☐ ☐ ☐ ☐ ☐ ☐ ☐ ☐

☐ ☐ ☐ ☐ ☐ ☐ ☐ ☐ ☐

☐ ☐ ☐ ☐ ☐ ☐ ☐ ☐ ☐ ☐ ☐

☐ ☐ ☐ ☐ ☐ ☐ ☐ ☐

☐ ☐ ☐ ☐ ☐ ☐ ☐ ☐

☐ ☐ ☐ ☐ ☐ ☐ ☐ ☐ ☐ ☐ ☐ ☐

Mean: _____

The **median** is the middle number in a set of data when the numbers are arranged in order.

Find the median, using the line plot.

4.

Ages of Students in the Photo Club	
Age	Students of Each Age
11	6
12	8
13	3
14	2
	19 ← Total

On the line plot count in order a number of Xs from the left and an equal number of Xs from the right to find the X in the middle. The middle X is above the column labeled 12. So the median age is 12.

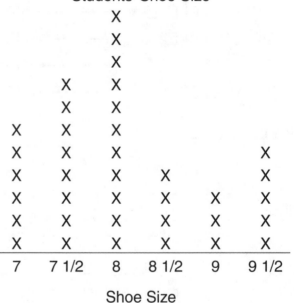

Ages of Students in the Photo Club

```
            X
            X
  X         X
  X         X  ← Median
  X         X
  X         X         X
  X         X         X         X
  X         X         X         X
 11        12        13        14
                 Age
```

The Xs represent the number of students of each age, or the **frequency**.

5.

Students' Shoe Size	
Shoe Size	Number of Students
7	6
$7\frac{1}{2}$	8
8	11
$8\frac{1}{2}$	4
9	3
$9\frac{1}{2}$	5
	37 ← Total

The middle X is above the column

labeled _____. A student with a

shoe size of $7\frac{1}{2}$ has a shoe size that

is *below* the median. A student with a

shoe size of $8\frac{1}{2}$ has a shoe size

_____ the median.

Students' Shoe Size

```
                X
                X
                X
        X       X
        X       X
  X     X       X
  X     X       X                   X
  X     X       X       X           X
  X     X       X       X     X     X
  X     X       X       X     X     X
  X     X       X       X     X     X
  7    7 1/2    8     8 1/2    9    9 1/2
              Shoe Size
```

Formulas and Variables

A **variable** is a symbol, such as a letter, that stands for possible numbers.
A **constant** is a specific number. Variables and constants, along with operations, make up expressions, equations, and formulas.

Name the variable(s) and constant(s) in each expression.

1. x + 5

letter — number

variable(s): _____ x _____

constant(s): _____ 5 _____

2. c + d

letter — letter

variable(s): _____

constant(s): _____ none _____

3. $2x$ + $2y$

number letter — number letter

variable(s): _____

constant(s): _____

4. $a \times a$

variable(s): _____

constant(s): _____

5. $\frac{t}{2}$

variable(s): _____

constant(s): _____

6. $6u + 8 - r$

variable(s): _____

constant(s): _____

Evaluate an expression by replacing all of the letters with numbers and computing the value.

Evaluate each expression. Complete the model to help you.

7. t − 8 for $t = 15$

☐☐☐☐☐☐☐☐☐☐☐☐☐☐☐

Cross out 8 tiles.

☐☐☐☐☐☐☐☒☒☒☒☒☒☒☒

There are 7 tiles left.

The value of $t - 8$ for $t = 15$ is _____ 7 _____.

8. 2 × ℓ for $\ell = 12$

☐☐☐☐☐☐☐☐☐☐☐☐

You need 2 groups of 12.

The value of $2 \times \ell$ for $\ell = 12$ is _____.

Find the value of each expression. Draw a model if it helps you.

9. $\frac{d}{t}$ for $d = 24$ and $t = 8$.

10. $a \times a$ for $a = 4$.

A **formula** is an equation that people have agreed to use to find needed amounts or measures. A formula contains variables and/or constants, operation(s), and an equals sign. The letters used for the variables often stand for words.

Name the variables and constants in each formula.

11. Perimeter = 4 × length of a side

This formula is used to find the perimeter of a square.

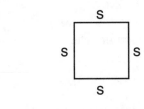

$$P \quad = \quad 4 \quad \times \quad s$$

letter   number   letter

variable(s): _____

constant(s): _____

12. Area = $\frac{1}{2}$ × base × height

This formula is used to find the area of a triangle.

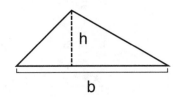

$$A \quad = \quad \frac{1}{2} \quad \times \quad b \quad \times \quad h$$

letter   number   letter   letter

variable(s): _____

constant(s): _____

When you **evaluate** a formula you replace all of the variables *except one* with numbers. The correct replacement for the remaining variable is the value you are looking for.

Evaluate each formula.

13. Area = base × height

This formula is used to find the area of a parallelogram.

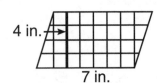

4 in.

7 in.

$A = b \times h$ for $b = 7$ in., $h = 4$ in.

$A = 7 \times 4$ Replace 2 of the 3 variables with numbers. You are looking for the area.

$A =$ _____ sq in.

14. Perimeter = 2 × length + 2 × width

This formula is used to find the perimeter of a rectangle.

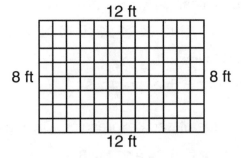

12 ft

8 ft   8 ft

12 ft

$P = 2 \times \ell + 2 \times w$ for $\ell = 12$ ft, $w = 8$ ft

$P =$ _____

$P =$ _____ sq ft

Percent Sense: Finding the Percent of a Number

Percent means per hundred. The symbol for percent is %. You can write a percent as a fraction, ratio, or decimal.

$25\% = \frac{1}{4} = 0.25$

$33\frac{1}{3}\% = \frac{1}{3} = 0.33\frac{1}{3}$

$50\% = \frac{1}{2} = 0.5$

$66\frac{2}{3}\% = \frac{2}{3} = 0.66\frac{2}{3}$

$75\% = \frac{3}{4} = 0.75$

$100\% = 1 = 1.0$

Choose one of the percents shown above, with its fraction and decimal equivalent, to describe the shaded area. Regardless of the size of the container, circle, segment, or other item, consider the whole as 100%.

1.
Half is shaded.

$50\% = \frac{1}{2} = 0.5$

2.

_____

3.

_____

4.

_____

5.

_____

6.

_____

7. $\underset{0 \quad 2 \quad 4 \quad 6}{\longmapsto}$

_____

8. $\underset{0 \quad 5 \quad 10}{\longmapsto}$

_____

9. $\underset{0 \quad 5 \quad 10 \quad 15 \quad 20}{\longmapsto}$

_____

Draw a model to show the given percent. Complete the equivalence.

10. 25%

$25\% = \frac{1}{2} = 0.5$

11. 50%

$50\% =$

12. 75%

$75\% =$

You can use grid paper and rectangles to find the percent of a number.

13. Find 25% of 8.

Draw a rectangle of 8 squares.

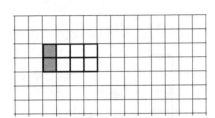

Shade $\frac{1}{4}$ of the 8 squares. The number of shaded squares is 25% of 8.

25% of 8 = ___2___

14. Find 25% of 20.

Draw a rectangle of 20 squares.

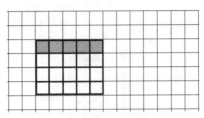

Shade $\frac{1}{4}$ of the 20 squares.
Count the shaded squares.

25% of 20 = _____

15. Find 25% of 4.

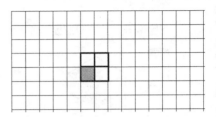

Shade $\frac{1}{4}$ of the 4 squares.

25% of 4 = _____

Shade the model to help you find the percent of the number.

16. 50% of 8

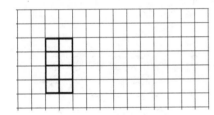

50% of 8 = _____

17. 50% of 16

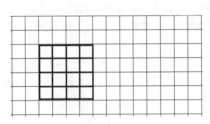

50% of 16 = _____

18. 50% of 2

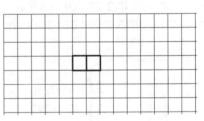

50% of 2 = _____

Find the percent of the number. Use line segments that begin at zero.

19. $33\frac{1}{3}$% of 9

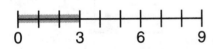

$\frac{1}{3}$ of the distance is shaded. The tick mark below the end of shaded part is 3.

$33\frac{1}{3}$% of 9 = 3

20. $33\frac{1}{3}$% of 12

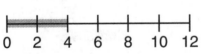

The tick mark below the end of shaded part is

_____.

$33\frac{1}{3}$% of 12 = _____

21. $33\frac{1}{3}$% of 15

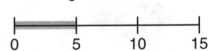

The tick mark below the end of shaded part is

_____.

$33\frac{1}{3}$% of 15 = _____

Adding and Subtracting Integers

You can add and subtract integers using tiles. Let ⊞ represent positive numbers and ⊟ represent negative numbers. If you pair the same number of ⊞ and ⊟ tiles, they cancel each other out, that is, their sum is 0. Because of this fact, ⊞ and ⊟ is called a **zero pair**.

Add. Use the tiles to help you.

1. $^+5 + {}^-3$

You have 3 zero pairs.
There are 2 ⊞ left.

$^+5 + {}^-3 = \underline{\ \ ^+2\ \ }$

2. $^-9 + {}^+4$

You have 4 zero pairs.
There are 5 ⊟ left.

$^-9 + {}^+4 = \underline{\hspace{1.5cm}}$

Draw a tile model to help you add.

3. $^-10 + {}^-7$

$^-10 + {}^-7 = \underline{\hspace{1.5cm}}$

4. $^+8 + {}^-11$

$^+8 + {}^-11 = \underline{\hspace{1.5cm}}$

Subtract. You can add zero pairs, because they equal 0.

5. $^-5 - {}^+3$

You have only ⊟s and you need to take away ⊞ ⊞ ⊞. Add 3 zero pairs.

Cross out 3 ⊞.
There 8 ⊟ left.

$^-5 - {}^+3 = \underline{\ \ ^-8\ \ }$

6. $^+6 - {}^-2$

⊞ ⊞ ⊞ ⊞ ⊞ ⊞

You have only ⊞s.
Add 2 zero pairs.

Cross out 2 ⊟.
There 8 ⊞ left.

$^+6 - {}^-2 = \underline{\hspace{1.5cm}}$

Add. Use a number line to help you.

7. ⁻9 + ⁻1

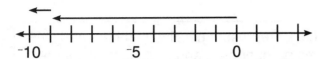

Start at 0 and move 9 spaces to the left, the negative direction. To add ⁻1 to ⁻9, move 1 space to the left, the negative direction.

⁻9 + ⁻1 = ___⁻10___

8. ⁻8 + ⁺5

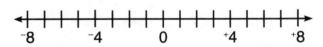

Start at 0 and move 8 spaces to the left. To add ⁺5 to ⁻8, move 5 spaces to the right, the positive direction.

⁻8 + ⁺5 = _____

Draw a number line to help you add.

9. 0 + ⁻7

0 + ⁻7 = _____

10. ⁻6 + ⁺6

⁻6 + ⁺6 = _____

To use a number line to subtract integers, count the spaces between the two integers. Then use the direction from the second integer to the first integer on the number line as the sign (⁺ or ⁻) of the difference.

Subtract. Use a number line to help you.

11. ⁻8 − ⁻4

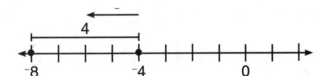

Locate ⁻8 and ⁻4 on a number line. There are 4 spaces between ⁻8 and ⁻4. From ⁻4 to ⁻8 move left, in the negative direction.

So ⁻8 − ⁻4 = ___⁻4___ .

12. ⁻7 − ⁺9

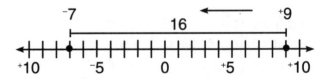

Locate ⁻7 and ⁺9 on a number line. There are 16 spaces between ⁻7 and ⁺9. From ⁺9 to ⁻7 move left, in the negative direction.

So ⁻7 − ⁺9 = _____ .

13. ⁺7 − ⁻5

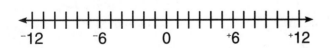

⁺7 − ⁻5 = _____ .

14. ⁺2 − ⁻6

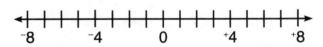

⁺2 − ⁻6 = _____ .

Comparing and Ordering Whole Numbers, Fractions, and Decimals

A number line can show whole numbers, decimals, and fractions in order from least to greatest. The greater number is always to the right.

On each number line write the missing numbers for A, B, C, and D.

1.

	A	0.5		B	1		D	1.5	C		2
0		$\frac{5}{10}$			$\frac{10}{10}$			$\frac{15}{10}$			$\frac{20}{10}$

2.

	0.25	C	A	1	B		D	2
0	$\frac{1}{4}$			$\frac{4}{4}$				$\frac{8}{4}$

3.

	$0.33\frac{1}{3}$	A	1		B	2	D	C	3
0	$\frac{1}{3}$		$\frac{3}{3}$			$\frac{6}{3}$			$\frac{9}{3}$

You can compare any two whole numbers, fractions, or decimals by graphing the numbers.

Compare. Write $<$, $=$, or $>$. Use a number line to help you.

4. $\frac{1}{4}$ _____ 0.5

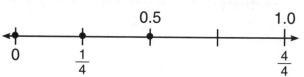

5. 0.75 _____ $\frac{1}{3}$

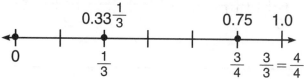

6. 0.2 _____ $\frac{1}{5}$

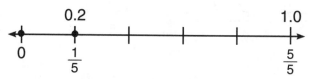

7. $2.33\frac{1}{3}$ _____ $1\frac{2}{3}$

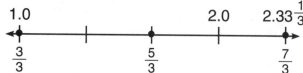

Compare. Write $<$, $=$, or $>$. Draw a number line to help you.

8. $\frac{3}{2}$ _____ 1.5

9. 1.25 _____ $1\frac{2}{10}$

Write the numbers in order from least to greatest. Use the number line to help you.

10. $\frac{1}{2}$, 0.1, 0.75

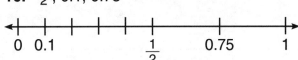

$\frac{1}{2}$ is to the right of 0.1 and 0.75 is to the right of $\frac{1}{2}$.

So, 0.1, $\frac{1}{2}$, 0.75 are in order from least to greatest — it is their order from *left* to *right* on the number line.

11. $\frac{1}{3}$, 1.3, $\frac{3}{2}$

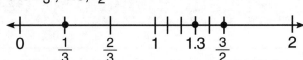

$\frac{1}{3}$

_____, _____, _____

12. 0.25, $\frac{3}{4}$, 0.6

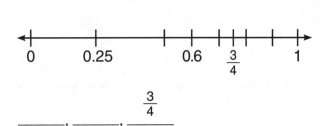

$\frac{3}{4}$

_____, _____, _____

13. 1.8, 0.8, $\frac{3}{5}$

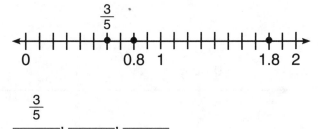

$\frac{3}{5}$

_____, _____, _____

Write the numbers in order from least to greatest.
Complete the number line to help you.

14. 0.4, 2.4, $1\frac{3}{4}$

_____

15. 1.2, 2.1, $2\frac{1}{2}$

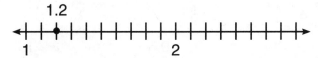

_____

Write the numbers in order. Draw a number line to help you.

16. 1.2, 1.25, $1\frac{9}{10}$

17. 1.75, $2\frac{3}{4}$, $2\frac{1}{4}$

Section III

Teacher's Edition
Intervention Workshop

Blackline Masters

Ones	ones														
	tens														
Thousands	hundreds														
	thousands														
	ten thousands														
Millions	hundred thousands														
	millions														
	ten millions														
Billions	hundred millions														
	billions														
	ten billions														
Trillions	hundred billions														
	trillions														
	ten trillions														
	hundred trillions														

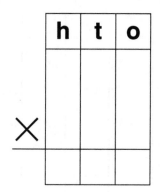

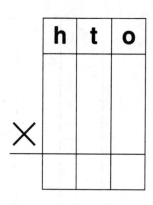

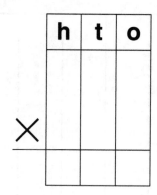

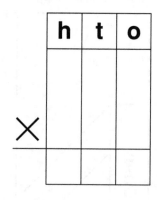

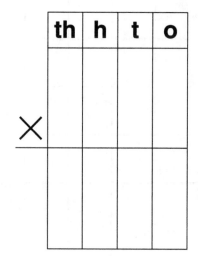

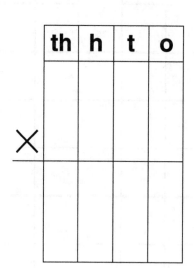

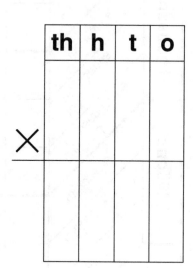

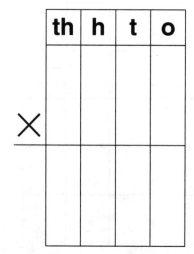

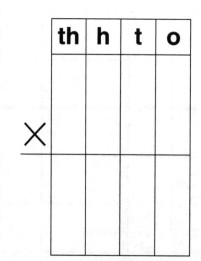

th	h	t	o

th	h	t	o

th	h	t	o

th	h	t	o

BLM 4

	hundreds	tens	ones$\bullet$	tenths	hundredths	thousandths	ten-thousandths	hundred-thousandths	millionths				
Ones													

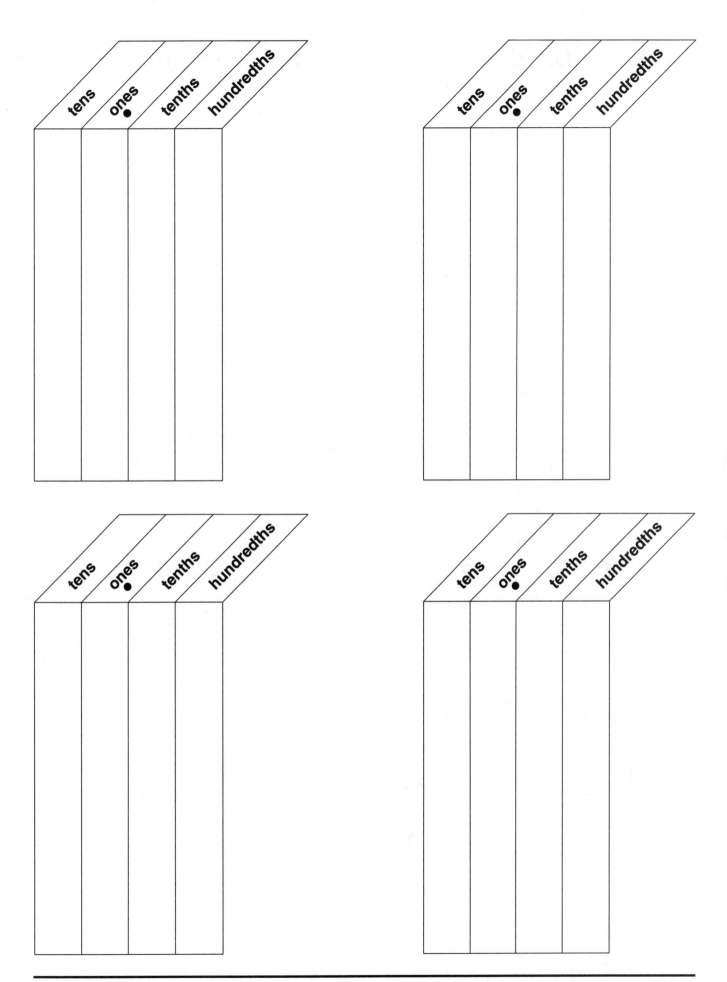

$	T	O.	d	p

$	T	O.	d	p

$	T	O.	d	p

$	T	O.	d	p

tens	ones •	tenths	hundredths	thousandths	ten-thousandths

tens	ones •	tenths	hundredths	thousandths	ten-thousandths

1

$\frac{1}{2}$	

$\frac{1}{4}$			

$\frac{1}{8}$							

$\frac{1}{5}$				

$\frac{1}{10}$									

$\frac{1}{3}$		

$\frac{1}{6}$					

$\frac{1}{9}$								

$\frac{1}{12}$											

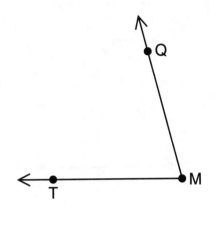

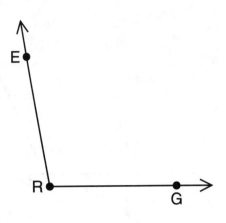

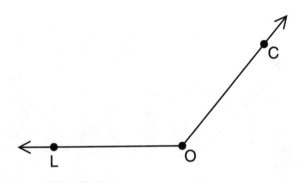

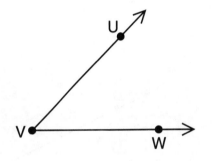

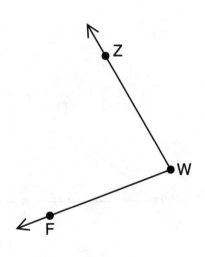

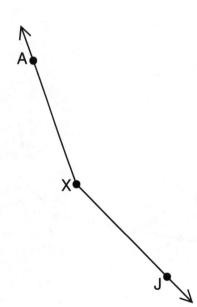

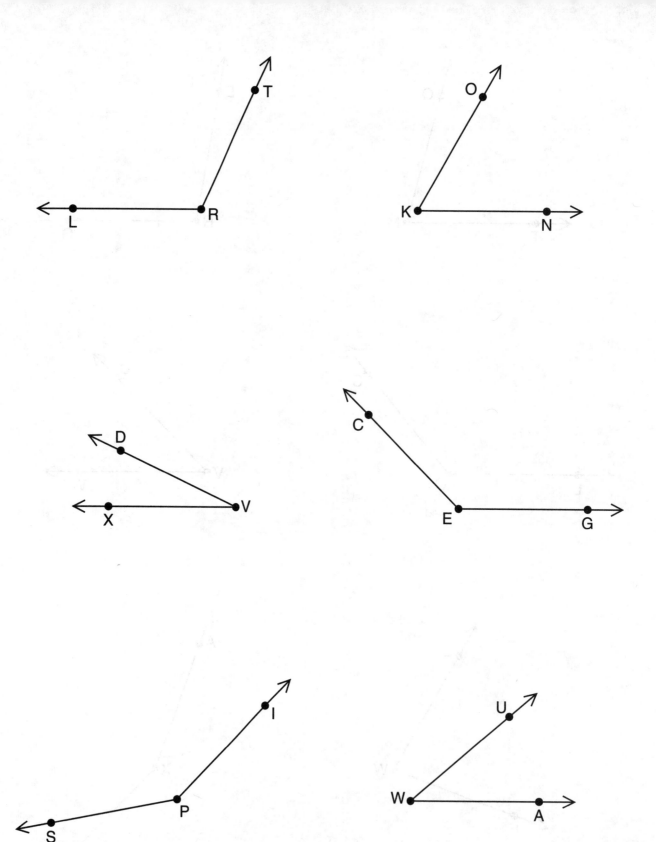

Copyright © by William H. Sadlier, Inc. Permission to duplicate classroom quantities granted to users of *Progress in Mathematics*.

Trapezoid	Parallelogram	Rectangle	Rhombus	Square
Exactly one pair of parallel sides	Two pairs of sides that are parallel and congruent	Two pairs of parallel congruent sides 4 right angles	Two pairs of parallel sides 4 congruent sides	Two pairs of parallel sides 4 congruent sides 4 right angles

Number of Sets	×	Number in Each Set	=	Number in All
	×		=	
	×		=	
	×		=	
	×		=	
	×		=	
	×		=	
	×		=	
	×		=	
	×		=	
	×		=	
	×		=	
	×		=	
	×		=	
	×		=	
	×		=	
	×		=	
	×		=	

Number in All	÷	Number in Each Set	=	Number of Sets
	÷		=	
	÷		=	
	÷		=	
	÷		=	
	÷		=	
	÷		=	
	÷		=	
	÷		=	
	÷		=	
	÷		=	
	÷		=	
	÷		=	
	÷		=	
	÷		=	
	÷		=	
	÷		=	
	÷		=	

Percent	Fraction
25%	
50%	
75%	
10%	
20%	
30%	
40%	
60%	
70%	
80%	
90%	
$33\frac{1}{3}\%$	
$66\frac{2}{3}\%$	

Rate percent (%)	of (✕)	Base total number	=	Percentage number that represents part of total
	✕		=	
	✕		=	
	✕		=	
	✕		=	
	✕		=	
	✕		=	
	✕		=	
	✕		=	
	✕		=	
	✕		=	
	✕		=	
	✕		=	
	✕		=	
	✕		=	
	✕		=	
	✕		=	
	✕		=	

Intervention Workshop Record Sheet

Name _____ Date _____

Diagnostic Interview Comments

_____

_____

_____

_____

Activity/Pupil Lesson Comments

_____

_____

_____

_____

_____

_____

Summing-It-Up-Quiz Comments

_____

_____

_____

_____

Summary

_____

_____

_____

_____

_____

Intervention Workshop Record Sheet

Name _____ Date _____

Diagnostic Interview Comments

_____
_____
_____
_____

Activity/Pupil Lesson Comments

_____
_____
_____
_____
_____
_____

Summing-It-Up-Quiz Comments

_____
_____
_____
_____

Summary

_____
_____
_____
_____
_____

Intervention Workshop Record Sheet

Name _____ Date _____

Diagnostic Interview Comments

_____

_____

_____

_____

Activity/Pupil Lesson Comments

_____

_____

_____

_____

_____

_____

Summing-It-Up-Quiz Comments

_____

_____

_____

_____

Summary

_____

_____

_____

_____

_____

Intervention Workshop Record Sheet

Name _____ Date _____

Diagnostic Interview Comments

_____
_____
_____
_____

Activity/Pupil Lesson Comments

_____
_____
_____
_____
_____
_____

Summing-It-Up-Quiz Comments

_____
_____
_____
_____

Summary

_____
_____
_____
_____
_____

Intervention Workshop Record Sheet

Name _____ Date _____

Diagnostic Interview Comments

_____

_____

_____

_____

Activity/Pupil Lesson Comments

_____

_____

_____

_____

_____

_____

_____

Summing-It-Up-Quiz Comments

_____

_____

_____

_____

Summary

_____

_____

_____

_____

_____

Intervention Workshop Record Sheet

Name _____ Date _____

Diagnostic Interview Comments

_____
_____
_____
_____

Activity/Pupil Lesson Comments

_____
_____
_____
_____
_____
_____
_____

Summing-It-Up-Quiz Comments

_____
_____
_____
_____

Summary

_____
_____
_____
_____
_____